Social

Social

Why Our Brains

Are Wired to Connect

Matthew D. Lieberman

B\D\W\Y
Broadway Books
New York

Published in the United States by Broadway Books, an imprint of
the Crown Publishing Group, a division of Random House LLC,
a Penguin Random House Company, New York.
www.crownpublishing.com

BROADWAY BOOKS and its logo, B \ D \ W \ Y, are trademarks
of Random House LLC.

Originally published in hardcover in the United States by Crown
Publishers, an imprint of the Crown Publishing Group, a division
of Random House LLC, New York, in 2013.

Library of Congress Cataloging-in-Publication Data is available
upon request.

ISBN 978-0-307-88910-2
eBook ISBN 978-0-307-88911-9

PRINTED IN THE UNITED STATES OF AMERICA

Book design by Lauren Dong
Illustrations credits: Fred Haynes (pages: 11, 22, 29, 42, 57, 67, 82, 166, 230,
and 245); Jane Whitney (pages: 17, 26, 51, 75, 113, 116, 135, 158, 171, 184,
185, 187, and 209)
Cover design by Oliver Munday

10   9   8   7   6   5

First Paperback Edition

For Naomi and Ian,
who showed me what my social brain was for

CONTENTS

Centuries ago, the philosopher Jeremy Bentham wrote, "Pain and pleasure . . . govern us in all we do, in all we say, in all we think." There is little doubt that we are drawn to physical pleasure and work hard to avoid physical pain. But do they "govern us in all we do"? Is this all that we are? I think they govern us far less than we typically assume. The institutions and incentive structures of society operate largely in accordance with Bentham's claim and thus are missing out on some of the most profound motivators of human behavior.

What Bentham and the rest of us typically overlook is that humans are wired with another set of interests that are just as basic as physical pain and pleasure. We are wired to be social. We are driven by deep motivations to stay connected with friends and family. We are naturally curious about what is going on in the minds of other people. And our identities are formed by the values lent to us from the groups we call our own. These connections lead to strange behaviors that violate our expectation of rational self-interest and make sense only if our social nature is taken as a starting point for who we are.

Over the past two decades, my colleagues and I have created a new kind of science called *social cognitive neuroscience*. Using tools like functional magnetic resonance imaging (fMRI), we have made startling discoveries of how the human brain responds to the social world—discoveries that were not possible before. These findings repeatedly reinforce the conclusion that our brains are wired to connect with other people. Some parts of the social mind can be traced back to the earliest mammals hundreds of millions of years ago. Other parts of the social mind evolved very recently and may be unique to humans. Understanding how these mental mechanisms drive our behavior is critical to improving the lives of individuals and organizations. This book will illuminate the neural mechanisms of the social mind and how they relate to making the most of our social lives.

Part One

Beginnings

CHAPTER 1

Who Are We?

Irv and Gloria lived the American dream for more than half a century. Depression-era children, they lifted themselves up from humble beginnings to become the toast of Atlantic City. They met when they were barely teens and spent their high school years going steady. Irv was admitted to Duke University, but then he signed up to serve his country as a naval pilot in World War II. When he went off to training camp, Gloria went with him. They were married just after the war and gave birth to two baby boomers who went on to become successful lawyers. Irv built the house he and Gloria lived in with his own hands. Later, he worked in real estate, and Gloria worked in his office with him. They had a knack for the business, and it didn't hurt that they had been savvy enough to purchase a few parking lots that the emerging casino industry would later want to snatch up. Irv and Gloria were inseparable. They lived, worked, and vacationed together.

At the age of sixty-seven, Irv learned that he had advanced prostate cancer, and he died soon after. Irv's death was a devastating blow to Gloria. People deal with tremendous adversity all the time and find ways to move on, but Gloria never did. She spent the rest of her days fixated on the loss of her partner, while her mind and memory slowly deteriorated. Over time, she became a different person. Before, she had always been charming and witty, if somewhat of a worrier. After Irv's passing, she became self-centered, inattentive, and even mean-spirited at times.

Gloria's friends wondered what had happened to her as they abandoned her one by one. Family struggled to put up with her moods and behavior. Most of the explanations offered for the changes she had undergone focused on neurobiology. Maybe she had some form of Alzheimer's disease or dementia? But nothing really supported such a diagnosis other than her growing memory loss. Some asked whether the medication she was taking to deal with her acute grief had left her with long-term neurological damage. Gloria, however, did not ponder such questions. She knew what was wrong—she would rather have died than live another day without Irv. I know this because she told me every chance she got. She was my grandmother. In her mind, she was dying of a broken heart. Years later, when I asked my father what had led her to change so radically, he said, "She died the moment he died. She didn't have a happy moment after."

Growing up, I had seen my grandparents as models of adulthood, of a strong, healthy marriage, and of the benefits of lifelong companionship. I spent my early summers living in their house, the one that Pop Irv built. I noticed how attentive and loving they were with each other and how they engaged with everyone else around them. Today, like Irv and Gloria, my wife and I work in the same profession in offices that are 20 feet apart. I learned from my grandparents that this is what it means to be happy. Why is it that the same relationship that can make you so happy for so many years can make life feel like it isn't worth living when the relationship is over or a loved one has passed on? Why have our brains been built to make us feel so much pain at the loss of a loved one? Could our capacity to feel so much pain be a design flaw in our neural architecture?

The research my wife and I have done over the past decade shows that this response, far from being an accident, is actually profoundly important to our survival. Our brains evolved to experience threats to our social connections in much the same way they experience physical pain. By activating the same neural circuitry that causes us to feel physical pain, our experience of social pain helps ensure the survival of our children by helping to keep them close to their parents. The neural link between social and physical pain also ensures that staying socially connected will be a lifelong need, like food and warmth. Given the fact that our brains treat social and physical pain similarly, should we as a society treat social pain differently than we do? We don't expect someone with a broken leg to "just get over it." And yet when it comes to the pain of social loss, this is a common response. The research that I and others have done using fMRI (functional magnetic resonance imaging) shows that how we experience social pain is at odds with our perception of ourselves. We intuitively believe social and physical pain are radically different kinds of experiences, yet the way our brains treat them suggests that they are more similar than we imagine.

Social will focus on three major adaptations in our brains that lead us to be more connected to the social world and better able to take advantage of these social connections to build more cohesive groups and organizations. The neural overlap between social and physical pain is the first of these adaptations. It ensures that we will spend our entire lives motivated by social *connection*.

Choosing a President

On October 21, 1984, President Ronald Reagan and his challenger, former Vice President Walter Mondale, held the second of two nationally televised presidential debates in the run-up to the presidential election. President Reagan remained popular, but his support was softening in light of growing concerns about his age. His poor performance in the previous debate, three weeks earlier, had opened the door to questions about his mental fitness. If reelected, Reagan would become the oldest sitting president in U.S. history (he was seventy-three at the time of the debate). Reagan's performance at this final debate is frequently cited as a turning point in the election, when Reagan's popular support solidified, contributing to the largest electoral landslide in history.

How did Reagan demonstrate that he was still in command of all of his faculties? Did he display his erudition on the current issues of the day? Did he play to his own strengths by vigorously attacking Mondale on issues like foreign policy or the tax code? No. It was Reagan's comedic timing that allowed him to carry the day. Reagan delivered a series of prefabricated one-liners with aplomb, regained his momentum, and never looked back. The most notable zinger came when the moderator asked him if age was a concern in the election. Reagan famously replied, "I will not make age an issue of this campaign. I am not going to exploit, for political purposes, my opponent's youth and inexperience." Mondale, not exactly a spring chicken at fifty-six, later commented that he knew at that very moment he had lost the campaign.

That night, nearly 70 million Americans watched the debate and came away convinced that the Gipper still had his mojo. Any fears people had that President Reagan had slipped were assuaged. But how we as a nation reached this conclusion on that night is surprising. Reagan himself didn't change our minds about him. It took a few hundred people in the audience to change our minds. It was their laughter coming over the airwaves that moved the needle on how we viewed Reagan.

Social psychologist Steve Fein asked people who had not seen the debate to watch a recording of it in one of two ways. Some individuals saw clips of the debate and the audience's reaction as it was played on live television, while others saw the debate without being able to hear the audience's reactions. In both cases, viewers heard the president deliver the same lines. Viewers who heard the audience laughter rated Reagan as having outperformed Mondale. However, those who did not hear the laughing responded quite differently; these viewers indicated a decisive victory for Vice President Mondale. In other words, we didn't think Reagan was funny because Reagan was funny. We thought Reagan was funny because a small group of strangers in the audience thought Reagan was funny. We were influenced by innocuous social cues.

Imagine watching the debate yourself (or maybe you did watch it). Would you think audience laughter could influence your evaluation of the candidates? Would you be influenced by those graphs that CNN shows at the bottom of the screen during today's debates to indicate how a handful of people are responding to the candidates, moment by moment? Would it sway your vote? Most of us, I suspect, would say no. The notion that our decision about who should be the president of our nation could be altered by the responses of a few people in the audience violates our theory of human nature, our sense of "who we are." We like to think of ourselves as independent-minded and immune to this sort of influence. Yet we would be wrong. Every day others influence us in countless ways that we do not recognize or appreciate. If this is true, why would our brains be built to be unwittingly influenced by people we don't even know?

Before judging the gullibility of our gray matter so harshly for using audience reactions to make sense of Reagan, let's take a moment to appreciate just how difficult it is to read other people's minds, to discern their character from the things they say and do. Thoughts, feelings, and personalities are invisible entities that can only be inferred, never seen. Assessing someone else's state of mind can be a herculean undertaking. Was Reagan still Reagan? Or had his mental faculties diminished? How could we know the difference without extensive neurological examinations? We all engage in this kind of mindreading of others every day; and it is so challenging that evolution gave us dedicated neural circuitry to do it.

While we tend to think it is our capacity for abstract reasoning that is responsible for *Homo sapiens'* dominating the planet, there is increasing evidence that our dominance as a species may be attributable to our ability to think socially. The greatest ideas almost always require teamwork to bring them to fruition; social reasoning is what allows us to build and maintain the social relationships and infrastructure needed for teams to thrive. That the brain has a network devoted to this kind of *mindreading* of others is the second of the three major brain adaptations I will discuss in this book.

The surprising thing is that even though social reasoning feels like other kinds of reasoning, the neural systems that handle social and nonsocial reasoning are quite distinct, and literally operate at odds with each other much of the time. In many situations, the more you turn on the brain network for nonsocial reasoning, the more you *turn off* the brain network for social reasoning. This antagonism between social and nonsocial thinking is really important because the more someone is focused on a problem, the more that person might be likely to alienate others around him or her who could help solve the problem. Effective nonsocial problem solving may interfere with the neural circuitry that promotes effective thinking about the group's needs.

The presence of a dedicated system for social reasoning in our brains still doesn't explain why most people watching the presidential debate were so affected by the responses of the audience. In this situation, the social reasoning system appears to have failed, resulting in distorted perceptions of the debate. Some part of our minds mistook anonymous audience laughter as a valid indicator of Reagan's mental vigor. Why would we substitute the judgment of others for our own? This was no momentary lapse. The world is filled with such laugh tracks and other contextual cues because our brains are designed to be influenced by others. Our brains are built to ensure that we will come to hold the beliefs and values of those around us.

In Eastern cultures, it is generally accepted that only by being sensitive to what others are thinking and doing can we successfully *harmonize* with one another so that we may achieve more together than we can as individuals. We might think that our beliefs and values are core parts of our identity, part of what makes us *us*. But, as I'll show, these beliefs and values are often smuggled into our minds without our realizing it.

In my research, I have found that the neural basis for our personal beliefs overlaps significantly with one of the regions of the brain primarily responsible for allowing other people's beliefs to influence our own. The self is more of a superhighway for social influence than it is the impenetrable private fortress we believe it to be. Our socially malleable sense of self, which often leads us to help others more than ourselves, is the third major adaptation I'll be discussing.

Social Networks for Social Networks

Most accounts of human nature ignore our sociality altogether. Ask people what makes us special and they will rattle off tried-and-true answers like "language," "reason," and "opposable thumbs." Yet the history of human sociality can be traced back at least as far as the first mammals more than 250 million years ago, when dinosaurs first roamed the planet. Our sociality is woven into a series of bets that evolution has laid down again and again throughout mammalian history. These bets come in the form of adaptations that are selected because they promote survival and reproduction. These adaptations intensify the bonds we feel with those around us and increase our capacity to predict what is going on in the minds of others so that we can better coordinate and cooperate with them. The pain of social loss and the ways that an audience's laughter can influence us are no accidents. To the extent that we can characterize evolution as designing our modern brains, this is what our brains were wired for: reaching out to and interacting with others. These are design features, not flaws. These social adaptations are central to making us the most successful species on earth.

Yet these social adaptations also keep us a mystery to ourselves. We have a massive blind spot for our own social wiring. We have a theory of "who we are," and this theory is wrong. The goal of this book is to get clear about "who we are" as social creatures and to reveal how a more accurate understanding of our social nature can improve our lives and our society.

Because real insight into our social nature has gained momentum only in the last few decades, there are tremendous inefficiencies in how institutions and organizations operate. Societal institutions are founded, implicitly or explicitly, on a worldview of how humans function. These are theories regarding the gears and levers of our nature that institutions try to operate on in order to strengthen society. Our schools, companies, sports teams, military, government, and health care institutions cannot reach their full potential while working from erroneous theories that characterize our social nature incorrectly.

The same holds true for teams within an organization. How should team leaders think about the social well-being of their team members? Does feeling socially connected make people socialize more and work less, or does it make team members work harder because they feel more responsibility for the team's success? Any team leader ought to know which of these claims is more likely to be true because it affects how the team should be managed. As we will see, neuroscience research indicates that ignoring social well-being is likely to harm team performance (and even individual health) for reasons we would not have guessed.

Just as there are multiple social networks on the Internet such as Facebook and Twitter, each with its own strengths, there are also multiple social networks in our brains, sets of brain regions that work together to promote our social well-being.

These networks each have their own strengths, and they have emerged at different points in our evolutionary history moving from vertebrates to mammals to primates to us, *Homo sapiens*. Additionally, these same evolutionary steps are recapitulated in the same order during childhood (see Figure 1.1). Parts Two, Three, and Four of this book each focus on one of these social adaptations:

- *Connection:* Long before there were any primates with a neocortex, mammals split off from other vertebrates and evolved the capacity to feel social pains and pleasures, forever linking our well-being to our social connectedness. Infants embody this deep need to stay connected, but it is present through our entire lives (Part Two: Chapters 3 and 4).

- *Mindreading:* Primates have developed an unparalleled ability to understand the actions and thoughts of those around them, enhancing their ability to stay connected and interact strategically. In the toddler years, forms of social thinking develop that outstrip those seen in the adults of any other species. This capacity allows humans to create groups that can implement nearly any idea and to anticipate the needs and wants of those around us, keeping our groups moving smoothly (Part Three: Chapters 5 through 7).

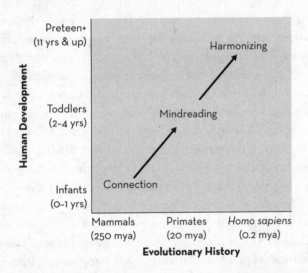

Figure 1.1 Emergence of Social Adaptations Across Evolution and Human Development (mya = millions of years ago)

- *Harmonizing:* The sense of self is one of the most recent evolutionary gifts we have received. Although the self may appear to be a mechanism for distinguishing us from others and perhaps accentuating our selfishness, the self actually operates as a powerful force for social cohesiveness. During the preteen and teenage years, adolescents focus on their selves and in the process become highly socialized by those around them. Whereas *connection* is about our desire to be social, *harmonizing* refers to the neural adaptations that allow group beliefs and values to influence our own (Part Four: Chapters 8 and 9).

Smarter, Happier, and More Productive

After considering how each of these networks shapes the social mind, we will turn to the all-important question for any scientific discovery: so what? How do we use what we have learned to improve the world in meaningful ways? In what ways do these social adaptations serve as principles for organizing groups, enhancing well-being, and bringing out the best in others and ourselves? In Part Five of the book, I will answer the *so-what* question for three domains of life. I will examine how our social connections can be enhanced in our daily lives to increase our overall well-being in life (Chapter 10). I will explore how we can make the workplace more responsive to our social wiring and how leaders can apply what we know about the social brain to improve workplace morale and productivity (Chapter 11). Finally, I will consider a number of ways that we can improve education, particularly in junior high, where motivation and engagement with learning typically plummet (Chapter 12). Humans are adapted to be highly social, but the organizations through which we live our lives are not adapted to us. We are square (social) pegs being forced into round (nonsocial) holes. Institutions often focus on IQ and income and miss out on the social factors that drive us. In Part Five, I will suggest ways to fix this, making us smarter, happier, and more productive. The social brain has a lot to teach us.

A Note

I came to the brain as an outsider, starting from an interest in philosophy and then getting a PhD in social psychology. I mention this here at the outset because I want you to understand that I appreciate what it is like to be interested in the brain but to find brain science daunting. The brain is at the seat of who we are, so it is intrinsically fascinating and holds the keys to unlocking untold mysteries. At the same time, the human brain is the most complicated device the universe has ever known. The brain contains billions of neurons, and each of these is connected to many others, creating an incalculable tangle of neural traffic. To make matters worse, we have awkward Latin names for each part of the brain (worse yet, multiple Latin names for the exact same part of the brain!). I spent years studying neuroscience before I stopped feeling completely overwhelmed. Throughout *Social*, I will focus on one brain region or system at a time. I will tell you what you need to know about the region or system, but I will keep the focus on what the study of such brain regions tells us about the mind, about who we are, about our social nature.

CHAPTER 2

The Brain's Passion

When I was in graduate school, I went through a rough breakup with a girlfriend that left me feeling lost, as if I were half a person. After a few months of self-pity and unhappy happy hours, I made a decision to devote myself to self-improvement. If I was half a person, I had room to develop the other half. So I set out to become who I wanted to be and who I thought I ought to be. Within a year, I forgot all about this self-improvement plan and was simply myself again, but for a year, I was devoted to this project of becoming.

During that year, I devoted a few hours each day to whatever it was I hoped would change my life for the better. But I had to make careful choices about how to spend these precious hours. I had to place my bets. What did I want to work at? What a person practices not only reveals what a person believes he or she can be better at but also reflects a bet that person has made about what is worth spending time improving. I decided to focus on becoming a better writer. I practiced writing in my spare time, throwing out entire passages I had written just to see if I could rewrite them more effectively. I studied art history, and I took folk guitar lessons, as well, but unlike my focus on writing, these pursuits didn't produce any ripples that still affect my life today.

It turns out our brains have a passion of their own; we know this because the brain seems to devote nearly all of its spare time to one thing. Unlike the different choices you and I might have made about how to divide up our free time, our brains, when given a chance, almost all seem to practice the same thing. Yes, our brains respond adaptively to whatever tasks they are given throughout the day. If you are an accountant completing a report on a deadline, the brain regions involved with math are recruited to support your calculations. If you are an art historian working as a curator at a museum, other brain regions might be brought online. But when the brain is not focused on a specific task, when there are no tax spreadsheets or art inventories to be updated, the brain turns to its lifelong passion.

What is it that the human brain likes to practice? Clearly, it must be extremely important to our success and well-being in life. The brain did not evolve over millions of years to spend its free time practicing something irrelevant to our lives. Indeed, the discovery that the brain is constantly practicing something suggests that evolution has, in a sense, made a bet about the value of that particular thing.

The Default Network

In 1997, Gordon Shulman and his colleagues at Washington University published two papers back to back in the same issue of the *Journal of Cognitive Neuroscience*, a prestigious journal for neuro-imaging research. At the time, positron emission tomography (PET) was a popular method for trying to identify which brain regions were involved in particular mental processes, such as memory, vision, and language. In PET scanning, subjects inhale radioactive tracers; using gamma rays, scientists can determine where blood, containing the tracers, is flowing to in the brain. When a region has more active neurons, more blood travels to that region. Prior to PET scanning, neuropsychologists were largely limited to waiting for the occasional and unfortunate brain injury as a result of disease or head trauma if they wanted to advance their understanding of where psychological processes occur in the brain. Sadly, the best periods in the history of neuropsychological research tended to cluster around major wars, because of all the war-related head wounds causing damage to different brain regions. PET scanning changed all of this. Scientists were able to study almost any psychological question, whenever they wanted, without harming anyone. It's hard to overstate how profound this advance was.

Shulman's two papers had a single mission: to look at a set of nine previous PET studies to determine whether there were brain regions that were always activated across a variety of mental activities studied by cognitive psychologists. The first of these papers examined which regions were commonly activated by different tasks, including motor, memory, and visual discrimination tasks (such as indicating when an image changed slightly). The results were a little disappointing: only a few regions showed increased activity across all the tasks, and they weren't very interesting brain regions. In hindsight, we know that these tasks rely on relatively distinct brain networks, so it makes sense that there wasn't much overlap across these tasks.

In the second paper the scientists asked the question, "What is more active in the brain when one is *not* doing one of these cognitive, motor, or visual tasks?" It was an unusual question. Typically neuroscientists have been interested in brain regions that are "switched on"—that become more active—when performing a task, identifying regions that help us accomplish it. Asking what in the brain becomes more active when you stop performing a task was a surprising approach. Thankfully, Shulman asked the question anyway. And he found a set of brain regions that were reliably more active when people were at rest, doing nothing, than when they were performing any of the specific tasks (see Figure 2.1). This paper set in motion a mystery that is still unsolved to this day. Why do we have brain regions that become more active when our minds go on their lunch break, so to speak—that is, when we are not doing anything in particular? It makes sense that areas of the brain involved in motor skills would quiet down when you finish doing a task that involves motor skills. But why would some regions of the brain systematically become *more* active when you are finishing a motor task—the same regions that become more active when you are finishing a visual task or a math problem?

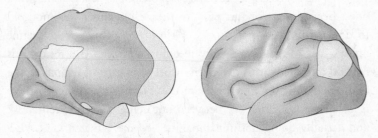

Figure 2.1 The Default Network

Calculatus Eliminatus

In the animated film of Dr. Seuss' *The Cat in the Hat,* a "moss-covered three-handled family gradunza" has gone missing. The Cat employs the official-sounding but entirely fictional technique of *calculatus eliminatus* in order to track it down. According to the Cat, *calculatus eliminatus* requires identifying all the places the missing object is not. The only remaining location will necessarily have to be where the missing object is. Hardly an efficient approach to locating where you left your car keys.

Nevertheless, early on, this was roughly the approach that scientists had to take with the network Shulman had discovered. Much more was known about what this network *did not do* than what it did. The early name given to describe the network was the "task-induced deactivation network" because it turned off in response to so many different kinds of tasks. In other words, tasks induce this network to turn off. Imagine having your job described in terms of all the things you *don't* do. You're the nonaccountant, nonmarketer, nonjournalist, nonsalesperson? Cool. But what is it you do exactly? The second name given to this network was the "default network" (or "default mode network"), which was better if only for its brevity. This name has stuck with neuroscientists. It refers to the fact that the network comes on by default when other tasks are finished.

Let's see if we can figure out a bit more about what this network does. Because participants lying in the PET scanner were not told to do anything during chunks of time when this network came on, it's easy to imagine they were doing nothing. As a result, it is natural to describe this default network as the set of regions in the brain that turn on when you are doing nothing. However, there is an enormous difference between not being given a specific task and actually doing nothing. Imagine you are lying inside a PET scanner. Let's say you are performing a routine cognitive task, such as indicating whether two letters on the screen are the same or different. After doing this for a minute, the word "Rest" appears. You know you have a minute of downtime before you have to start performing the boring task again.

The experimenter can't gauge what you do next, but your mind is hardly at rest. Go ahead and close your eyes for thirty seconds and try. If you did, your mind probably darted around from one thought, feeling, or image to another. Instead of being at rest, your mind was highly active. If you are like most people, you thought about other people, yourself, or both. In other words, you engaged in what psychologists call *social cognition,* which is simply another way of describing thinking about other people, oneself, and the relation of oneself to other people. A college sophomore asked to do boring repetitive tasks in a psychology experiment in order to earn money to take someone on a date will start thinking, as soon as there is a break in the task, about the girl, the date, and whether or not she really likes him.

So perhaps the default network that comes on when we are given a break from performing cognitive tasks is involved in social cognition, the capacity to think about other people and ourselves. It took a while to find out whether or not that was true because social neuroscientists weren't paying attention to research on the default network at first. What the brain does when we stop doing a motor task does not sound like the kind of thing a social neuroscientist would usually care about. But as it happens, the network in the brain that reliably shows up during social cognition studies is virtually identical to the default network. In other words, the default network supports social cognition—making sense of other people and ourselves.

Default Social Cognition

At this point, you might think, "Isn't it obvious that people think about people when they are not otherwise engaged? Why is that so interesting?" When I first noticed the overlap between the default network and the social cognition network, I didn't think it was particularly significant for this very reason. All this overlap tells us is that people typically have a strong interest in the social world and are likely to choose to think about it when they have free time.

I have since become convinced that I had the relationship between these two networks backward. And this reversal is tremendously important. Initially, I thought, "We turn on the default network during our free time because we are interested in the social world." While that is true, the reverse is also true and far more interesting: I now believe "we are interested in the social world *because we are built to turn on the default network during our free time.*" In other words, if this network comes on like a reflex, it may nudge our attention toward the social world. And not just to other people as objects in our environment. Rather, the default network directs us to think about other people's minds—their thoughts, feelings, and goals. To take what philosopher Daniel Dennett called "the intentional stance," it promotes understanding and empathy, cooperation, and consideration. It suggests that evolution, figuratively speaking, made a big bet on the importance of developing and using our social intelligence for the overall success of our species by focusing the brain's free time on it. I bet a year on becoming a better writer; evolution bet millions of years on making us more social.

But is there any reason to believe this claim that default network activity can be a cause, rather than a consequence, of our interest in the social world? Is there evidence that it is a leading, rather than a lagging, indicator of social thinking? There are a few provocative findings that suggest default network activity during rest may reflect an evolved predisposition to think about the social world in our free time rather than its being merely a moment-by-moment personal choice.

One key finding comes from newborns. Babies show default network activity almost from the moment of birth. One study looked at which brain regions were engaged in highly coordinated activity in two-week-old babies and found that the default network was chugging away just as it does in adults. Another group found evidence of a functional default network in two-day-old infants. However, the same pattern was not seen in infants born prematurely, suggesting that this mechanism is engaged and set to turn on when we are most likely to enter the social world.

Why does the presence of default network activity in infants matter? Because infants clearly haven't cultivated an interest in the social world yet, or in model trains, or in anything. Two-day-old infants cannot even focus their eyes yet. In other words, the default network activity precedes any conscious interest in the social world, suggesting it might be instrumental in creating those interests.

You might be familiar with the claim that Malcolm Gladwell made famous in his book *Outliers* that it takes 10,000 hours of practice to become an expert at something. Although different people might put those 10,000 hours toward becoming a concert violinist, professional athlete, or Xbox superstar, the brain puts in the 10,000 hours and more to enable us to become experts in the social world. One study found that 70 percent of the content in our conversations is social in nature. Assuming that we spend just

20 percent of our time in general thinking about other people and ourselves in relation to others, our default network would be engaged at least three hours a day. In other words, our brains have put in 10,000 hours before we reach the age of ten. The repeated return of the brain to this social cognitive mode of engagement is perfectly situated to help us to become experts in the enormously complex realm of social living.

There is a second reason to think default network activity is often a cause, rather than a consequence, of our focus on the social world. Typically, the default network is studied by giving people extended periods of rest, ranging from thirty seconds to several minutes. It is easy to imagine that with all that time, people intentionally turn their minds to whatever matters to them in their daily lives. But what if people had only a few seconds of downtime? Imagine solving a math problem; afterward, you know you have just two seconds before the next math problem. It's unlikely that people would decide to try to think about anything other than getting ready for the next math problem. Nevertheless, when Robert Spunt, Meghan Meyer, and I gave people only a few seconds of pause between math problems, they showed almost the same default network activity as when they had much longer breaks. In fact, the default network activity was present the instant the math problems were finished. This suggests that the default network really does come on like a reflex. It is the brain's preferred state of being, one that it returns to literally the second it has a chance.

In psychology, *priming* refers to seeing or thinking of something that prepares you to do something more efficiently right after. Consider what happens when you read the word "face." Now turn to the next page, and look at Figure 2.2. What do you see? You are more likely to see faces at first because seeing the word "face" primed you. It prepared your brain to see faces. As we will see in Chapter 5, there are now data suggesting that the brain's reliable and rapid return to its default state similarly serves to prime us to be prepared for effective social thinking.

Figure 2.2 Rubin's Illusion.
Adapted from Rubin, E. (1915/1958). Figure and ground. In D. C. Beardslee &
M. Wertheimer (Eds.). Readings in Perception. Princeton, NJ: Van Nostrand, pp. 194–203.

The default network quiets down when we perform a specific task, such as calculating a math problem in math class or studying ancient Greek pottery in history class. But when the mind's chores are done, it returns to Old Faithful—the default mode. In other words, the brain's free time is devoted to thinking socially. Consciously or not, it seems to be processing (and perhaps reprocessing) social information, as well as priming us for social life. It might be using this time to integrate new experiences into our long-standing knowledge of other people, their relationships with one another, or our relationships with them. It might be used to extract information from recent interactions to update the general rules we use for understanding the minds of others. This neural habit is at work in two-day-old infants and in our adult brains the moment we stop whatever else we are doing. In essence, *our brains are built to practice thinking about the social world and our place in it.*

If the brain practices thinking socially from infancy through adulthood, the implication is that evolution has made a major bet on the value of our becoming social experts, and in our being prepared in any given moment to think and behave socially. This constant practice doesn't mean we have perfected being social. We haven't. But without this practice, think how much worse-off we might be. There are so many other things our brain could have been built to spend its spare time on—learning calculus, improving our logical reasoning ability, cataloging variations in the classes of objects we have seen. Any of these could have adaptive value. But evolution placed its bet on our thinking socially.

Accidentally Social?

The popular conception of human nature emerging from psychology over the last century suggests that we are something of a hybrid, combining reptilian, instinct-driven motivational tendencies with superior higher-level analytic powers. Our motivational tendencies evolved from our reptilian brains eons ago and focus on the four Fs: fighting, fleeing, feeding, and fooling around. In contrast, our intellectual capacities are relatively recent advances. They are what makes us special.

One of the things that distinguish primates from other animals, and humans from other primates, is the size of our brains—in particular, the size of our *prefrontal cortex*, that is, the front part of the brain sitting right behind the eyes. Our big brains allow us to engage in all sorts of intelligent activities. But that doesn't mean our brains evolved to do those particular things. Humans are the only animals that can learn to play chess, but no one would argue that the prefrontal cortex evolved specifically so that we could play the game of kings. Rather, the prefrontal cortex is often thought of as an all-purpose computer; we can load it up with almost any software (that is, teach it things). Thus, the prefrontal cortex seems to have evolved for solving novel hard problems, with chess being just one of an endless string of problems it can solve.

From this perspective there might not be anything special at all about our ability and tendency to think about the social world. Other people can be thought of as a series of hard problems to be solved because they stand between us and our reptilian desires. Just as our prefrontal cortex can allow us to master the game of chess, the same reasoning suggests that our all-purpose prefrontal cortex can learn to master the social game of chess—that is, the moves that are permissible and advantageous in social life. From this perspective, intelligence is intelligence whether it's being applied to social life, chess, or studying for a final exam. The creator of one of the most widely used intelligence tests espoused this view, arguing that social intelligence is just "general intelligence applied to social situations." This view implies social intelligence isn't special and our interest in the social world is just an accident—a consequence of the particular problems we are confronted with.

One of the standards we can apply to determine whether a human characteristic is accidental or not is its universality. I would guess that less than 10 percent of the world's population plays baseball, making it a good candidate for an accidental ability. Almost everyone could learn to play, but few do. In contrast, standing upright is a human universal. Learning language of some kind is nearly universal. So is reasonably good eyesight. In a study of more than 13,000 people, 93 percent had good eyesight. As a back-of-the-envelope calculation, 93 percent seems like a reasonable benchmark to say something might have been significant enough in its own right to promote evolutionary adaptation.

From that perspective, can we conclude our sociality is an accident if more than 95 percent of people report having friends? If you take an alien's-eye view, friendship is a quirky phenomenon. Every friend begins as a stranger to us, typically someone we share no genes with, possibly representing an unknown threat. And yet this person may be someone we ultimately choose to disclose our innermost secrets and vulnerabilities to, or depend on more than anyone else in the world. Friendship has been documented in only a few species, but it is nearly universal in humans. Perhaps we can acquire more resources if we have friends. Perhaps they can be seen as a means to an end. If so, we should keep track in any friendship of how much we give and receive in order to ensure that we are getting our due (and hopefully more). Yet the closer friends become, the less they tend to keep track of who has done more or less for one another. Often, a friend's primary value is the comfort of knowing we have friends. Despite the various ways friends can be directly useful to us, the fact that our friends are our friends is often an end in itself.

And then consider Facebook. There are more than a billion people with Facebook accounts. Facebook is the most commonly visited website in the world, ahead of Google, Yahoo!, eBay, and Craigslist. The Internet dominates our lives as no technology has before. And the place we go to most often is Facebook. That's because Facebook offers the best deals on . . . nothing. If Facebook were a religion (and some argue that it is), it would be the world's third largest behind Christianity (2.1 billion) and Islam (1.5 billion). Americans spend 84 billion minutes per month engaged in religious activities—and 56 billion minutes on Facebook.

What Facebook does provide is an efficient way to stay connected with the people in our lives. It allows us to keep in touch with people we don't get to see as often as we want or to reconnect with people from our past or to relive the fun of last night's party with all our friends who were there. Is it just an accident that the single most successful destination on the Internet, or anywhere else, is a place entirely dedicated to our social lives?

If our sociality were an accident, simply another use of our big brains to achieve our selfish ends by manipulating others, would we altruistically help others in need whom we will never meet, who will never know of our good deeds? We give to others for many reasons, but one reason is that we are wired to feel empathy and compassion for the plight of others. When we see others in need, at least some of the time we think, "Something must be done." Apparently this kind of compassion happens quite a bit. In the United States alone, we give an average of $300 billion a year to charities worldwide. That is an awfully big accident.

If social intelligence were a random application of our general intelligence, we would expect to see the same brain regions associated with both kinds of intelligence. That would be a sensible story if it were true, but it isn't. The brain regions reliably associated with general intelligence and its related cognitive abilities, like working memory and reasoning, tend to be on the outer (or *lateral*) surface of the brain (see Figure 2.3), whereas thinking about other people and oneself utilizes mostly *medial* (or *midline*) regions of the brain (see Figure 2.1).

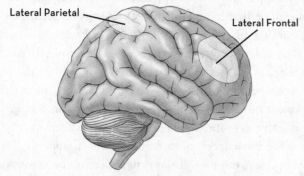

Figure 2.3 Brain Regions Associated with Working Memory in the Lateral Frontal and the Parietal Regions

Moreover, neural networks that support social and nonsocial thinking often work at cross-purposes—much like the two ends of a *neural seesaw*. If we look at the brain when a person isn't being asked to do anything in particular, we see the social cognition network turned on. Typically, the more this network turns on, the more the general cognition network responsible for other nonsocial kinds of thinking turns off. Likewise, when people engage in nonsocial thinking, the general cognition network turns on and the social cognition network turns off. (I am using "turn on" and "turn off" colloquially. Brain regions do not actually turn off. Rather, they become less active under some conditions and more active under others.) To the extent that the social cognition network stays on when we engage in nonsocial thinking, it tends to interfere with our ability to perform. This is hard to reconcile with the idea that the prefrontal cortex is an all-purpose computer that uses the same random-access memory (RAM) chips to think about office politics as it uses to play chess and figure out our taxes.

Part of what makes it hard to believe that social cognition and nonsocial cognition depend on different neural machinery is that these two kinds of thinking don't *feel* very different when we use one versus the other. It's not like the change we feel when speaking in our native tongue compared with speaking in a recently learned language. It's not like the distinct experiences we have when solving a math problem and when imagining being a superhero flying through the air. These differences feel really different to us. But when we switch from thinking socially to thinking nonsocially, we feel as if we have simply changed topics, rather than changing the way in which we are thinking. But that does not mean the differences between social and nonsocial thinking aren't real. It means only that the differences aren't conspicuous to us.

We do have at least one way of intuitively appreciating differences between social and nonsocial thinking. Most of us subscribe to the common wisdom that book smarts and social smarts rarely go together. These two kinds of intelligence seem to require different abilities, and the brain has separate networks to support them. A recent study of children with Asperger's disorder brings this distinction home. Asperger's is considered to be a milder version of autism, but it is associated with many of the same deficits in social cognition and social behavior. A group of children with Asperger's actually performed better on a test of abstract reasoning than age-matched healthy children. If social intelligence and nonsocial intelligence compete with each other, like the two ends of a seesaw, then it makes sense that deficits that take away some of the strength and power on one end of the seesaw will give the other end greater influence.

Bigger Brains

Most of us have been taught that our bigger brains evolved to enable us to do abstract reasoning, which promoted agriculture, mathematics, and engineering as complex tools to solve the basic problems of survival. But increasing evidence suggests that one of the primary drivers behind our brains becoming enlarged was to facilitate our social cognitive skills—our ability to interact and get along well with others. All these years, we've assumed the smartest among us have particularly strong analytical skills. But from an evolutionary perspective, perhaps the smartest among us are actually those with the best social skills.

Before we discuss the reasons why the human brain is larger, we need to know what it means to say the human brain is larger than the brains of other species. There are countless ways that brains can be compared to one another—total volume, weight, number of neurons, degree of cortical convolution, total gray matter volume, and total white matter volume. And those are just the tip of the iceberg.

One important preliminary fact is that brain size is predicted very well by body size. This means that a great deal of absolute brain size is associated with things like maintaining and monitoring the body. The bigger the body, the more brain tissue is needed to oversee it. As a result, really big animals tend to have really big brains. Indeed, if only brain weight is considered, humans are nowhere near the top of the heap. The human brain weighs in at about 1,300 grams, just about equal to the brain of the bottlenose dolphin. African elephants' brains nearly triple that at 4,200 grams, and some whales have brains that can reach 9,000 grams. Humans do better comparatively when we consider the total number of neurons in the brain. We have approximately 11.5 billion neurons, which is the highest known number in the animal kingdom . . . but just barely. Killer whales have 11 billion neurons. If intelligence were only a matter of the number of neurons we possessed, we would be building eighty-story sky-scrapers, and killer whales would be building seventy-five-story seascrapers.

Despite the strong relationship between body size and brain size, some animals have larger brains than their body size would seem to require for the basic maintenance and monitoring functions. The degree to which an animal's brain size deviates from what we would expect, based on body size, is called *encephalization*. It is thought to represent the brain's spare capacity to do more than control the body—like developing intelligence. Here, humans are the undisputed heavyweight champions of the animal kingdom. Human encephalization is 50 percent greater than that of the next closest animal, the bottlenose dolphin, and nearly twice that of any nonhuman primate (see Figure 2.4). And, just as we would expect, newer parts of the brain, like the prefrontal cortex, show this enhanced encephalization as well.

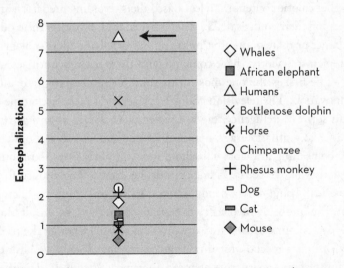

Figure 2.4 Encephalization Across Species. Arrow points to humans.
Adapted from Roth, G., & Dicke, U. (2005). Evolution of the brain and intelligence. *Trends in Cognitive Sciences,* 9(5), 250–257.

Making MacGyvers?

So, why did the human brain become so much larger, in terms of encephalization, than the brains of other animals? Making a bigger brain comes at a great cost in an animal's time and energy. It is not much of an exaggeration to say we live in order to feed our brains. In adult humans, the brain makes up approximately 2 percent of the total body mass, and yet it consumes (that is, metabolizes) 20 percent of its energy. In prenatal infants, the brain consumes 60 percent of the body's total metabolism, a rate that continues through the first year of life and only gradually declines to the 20 percent level during childhood.

The brain's outsized energy budget means that evolution would have selected for brain growth only if brain growth helped primates solve problems critical to survival and reproduction. Such problems include finding and extracting foods like fruit and meats with higher calorie content than leaf-based diets, avoiding predators, and keeping their young safe. So what particular kind of cleverness does a larger primate brain offer in the service of solving these ecological problems? Scientists have come up with three main hypotheses.

The first is the one most of us think of intuitively: individual innovation. The television character MacGyver is the archetype of this sort of intelligence. He is a secret agent who is always getting into sticky situations and manages to innovate his way out of them by combining available household items in novel ways to produce exactly what he needs. In one episode, he stops a dangerous sulfuric acid leak using only a candy bar and its tin foil wrapper. Though our lives probably involve fewer explosive situations, we are all MacGyvers in our own way. We are problem solvers, whether the problem is what to cook for dinner given the ingredients we have on hand or how to structure a spreadsheet most effectively. To differing degrees, all primates are problem solvers. When we think about having big brains, we think about how smart they make us, as individuals, at learning and solving problems. But despite this being an obvious answer—and perhaps the answer you might have been taught in high school science—it's not the right answer. Whether a species innovates more or less is not the best predictor of brain size across species.

The second hypothesis focuses on our social abilities. Although humans as a species are very good at inventing solutions to problems, individuals don't always do as well on their own. When my son, Ian, was four, he loved to play the videogame *Super Hero Squad*. My wife, Naomi, and I would always have to play with him because he would get stuck very easily. The game involved solving a series of puzzles, and Ian wasn't old enough to solve them on his own. Out of every five puzzles in the game, Ian might have been able to solve one. Of the same five puzzles, Naomi and I could solve only two or three. Generating the solutions was just too hard. Apparently, we were too old to solve the problems because the way we would move forward is by watching YouTube "walkthroughs" of a young boy getting through the puzzles successfully and explaining the tricks as he went.

In other words, humans don't excel as a species because we are all innovators. Rather, one or more of us (in this case, a young videogame wizard) devise a solution to a common problem, and the rest of us learn the solution from that person by imitation or instruction. Perhaps we developed larger brains to improve our capacity for imitation or social learning? While species that engage in social learning more often do have bigger brains, it turns out that this is not the best predictor of brain size across species either.

The Social Brain Hypothesis

The third hypothesis for why we have bigger brains suggests that we have them so that we can connect and cooperate with one another. If you needed to build a home on your own, how well would you do? Could you build yourself a log cabin? Cutting and lifting logs is a lot easier with a couple extra pairs of hands. In a sense, the basis of society could be seen as an agreement that if you help me build my log cabin, I'll help you build yours in turn. Everyone gets a better home, and we all benefit. Nonhuman primates aren't in the business of building log cabins, but their success at dealing with ecological problems can also be profoundly improved by dealing with their problems together—through coordinated cooperative action. Survival turns out not to be a zero-sum game for primates.

In the early 1990s, evolutionary anthropologist Robin Dunbar made the provocative claim that the primary reason the neocortex grew larger was so that primates could live in larger groups and be more actively social. *Neocortex ratio* refers to the size of the neocortex relative to the size of the rest of the brain. The evidence that Dunbar and others have marshaled is impressive. When the relative size of the neocortex is correlated with differences in the three potential drivers of brain size (individual innovation, social learning, and group size), group size is the strongest predictor of neocortex size. In his first study, Dunbar pitted group size against indicators of nonsocial kinds of intelligence, and he found that although both correlated with the neocortex ratio, group size was the better predictor. Later work demonstrated that these effects were strongest among the frontal lobes.

Using the equations that emerged from this line of work, Dunbar was able to estimate what the largest effective, coherent social group should be for each kind of primate, based on its neocortex ratio. His analysis suggests that for humans the number is around 150, the largest for any primate. This is referred to as "Dunbar's number," and it turns out that a striking number of human organizations tend to operate at around that size. For instance, village size, estimated from as long ago as 6000 BC and as recently as the 1700s, converges around the 150 mark. Ancient and modern armies also organize around units of about 150 people.

The human brain didn't get larger in order to make more Mac-Gyvers. Instead, it got larger so that after watching an episode of *MacGyver*, we would want to get together with other people and talk about it. Our social nature is not an accident of having a larger brain. Rather, the value of increasing our sociality is a major reason for why we evolved to have a larger brain.

Making Groups Worth It

What is so beneficial about living in larger groups? Why would evolution foster an increase in our typical group size by increasing the size of our brains? The most obvious advantage to larger groups is that predators can be strategically avoided or dealt with more successfully. It's hard to keep your mind focused on finding food when you are worried about being food, and it's dangerous to be out in the open looking for food by yourself. Groups of apes, in contrast, can trade off time looking for food and watching out for predators. That is a big advantage.

The downside of larger groups is that there is increased competition for food and mating partners within the group. If you are on your own and you manage to find food, it's yours. The larger your group, the more likely it is that one of the others in your group will try to poach it. Primates with strong social skills can limit this downside by forming alliances and friendships with others in their group.

Consider two chimpanzees, Smith and Johnson. Johnson gets bullied regularly by Smith. Johnson is a relatively low status ape. But if he can form an alliance with Brown, a high status ape, this will help protect him from Smith. Because Brown is high status, he knows that if he takes Johnson's side in a skirmish with Smith, Smith will stand down immediately, not wanting take on a higher status chimp. This is a great deal for high status Brown because he will get more favors (for example, grooming) from his low status partner, Johnson, without really putting himself at risk in confrontations with Smith.

Even for chimpanzees, there are a lot of social dynamics at work here. For Smith, Johnson, and Brown to form the alliances that work best for each of them, they need to keep track of a great deal of social information. They need to keep track of everyone's status relative to themselves, but they also need to know the status of each chimpanzee relative to the others. If there are just 5 chimps in a group, each chimp needs to keep track of the social dynamics of 10 chimp-to-chimp relationships, or *dyads*. A group of 15 requires keeping track of 100 chimp-to-chimp relationships to be fully informed. Triple the group size to 45, and now there are 1,000 dyadic relationships. When we reach Dunbar's number, a group of 150 individuals, there are more than 10,000 possible relationship pairs to consider. So we can begin to see why a bigger brain might come in handy. While there is a tremendous upside to being part of a group, that is true only if you know how to play the odds and form the right coalitions to avoid the downsides of group living. It requires an expansive capacity for social knowledge.

The same is true, of course, for humans. To give an example, every year, thousands of undergraduates apply to the most prestigious PhD programs in the United States. A big part of getting in is having persuasive letters of recommendation that are sent on one's behalf. These letters have been subjected to the same grade inflation forces running rampant across college campuses. Thus the assessments typically range from "this is a fantastic student" to "this is the *most* fantastic student." When I read these letters, what often matters more to me than the content of the letters is who wrote them. When a fellow social or affective neuroscientist writes a glowing letter, it is highly meaningful to me because that person is accountable to me the next time we are at a conference together. In contrast, professors in anthropology can write a glowing letter to me with impunity, regardless of any flaws in the candidate, because

I probably don't know them and they won't be held accountable. For this very reason, their letters do not hold as much weight with me. The upshot of all this is that a college sophomore thinking about which lab to volunteer for will tangibly benefit from knowing how a potential mentor in the department is viewed by the professors the student might want to study under to get a PhD a few years later. This is complex social cognition.

Many of the important innovations created by human beings—steam engines, lightbulbs, and X-rays—were created by a few individuals whose work was shared with the world at large. The majority of human beings would not have come up with these solutions in a hundred lifetimes. I know I wouldn't have. Most of us create very little that advances civilization. But each of us needs to navigate complex social networks to be successful in our personal and our professional lives. Primate brains have gotten larger in order to have more brain tissue devoted to solving these social problems, so that we can reap the benefits of group living while limiting the costs.

Part Two

Connection

CHAPTER 3

Broken Hearts and Broken Legs

Comedian Jerry Seinfeld used to tell the following joke: "According to most studies, people's number one fear is public speaking. Death is number two. Does this sound right? This means to the average person, if you go to a funeral, you're better-off in the casket than doing the eulogy." The joke is a riff based on a privately conducted survey of 2,500 people in 1973 in which 41 percent of respondents indicated that they feared public speaking and only 19 percent indicated that they feared death. While this improbable ordering has not been replicated in most other surveys, public speaking is typically high on the list of our deepest fears. "Top ten" lists of our fears usually fall into three categories: things associated with great physical harm or death, the death or loss of loved ones, and speaking in public.

Of course our fear of physical harm is precisely why we evolved an experience of fear in the first place. Would-be ancestors who lacked a basic fear of dangerous threats probably never became our ancestors because they did not live long enough to reproduce. Fearing the loss of loved ones makes evolutionary sense too because they help pass on our genes. But public speaking? Darwin didn't have a lot to say about that one because there is no obvious connection between public speaking and survival. So what are we afraid of when we think about speaking in public? We all speak, and most of us are quite comfortable speaking with friends, family, and colleagues. So it isn't speaking per se that gives us butterflies. It's the public part of public speaking that terrifies so many of us—whether it's speaking in front of a dozen, a hundred, or a thousand strangers.

You may have seen some of the same after-school television specials I did growing up. The sixth grader gets up to give a speech in front of an auditorium filled with other kids. He flubs his lines and becomes the laughingstock of the school (until he does something unexpectedly brave and wins the heart of the cutest girl in school). I suspect most of us have a fear that parallels this scene. We are afraid that everyone will think we are foolish or incompetent. We are afraid that everyone will reject us. Indeed, speaking in front of a large audience probably maximizes the number of people who could all reject us at one time.

What is curious is that the person speaking probably doesn't know or care about most of the people there. So why does it matter so much what they think? The answer is that it hurts to be rejected. Ask yourself what have been the one or two most painful experiences of your life. Did you think of the physical pain of a broken leg or a really bad fall? My guess is that at least one of your most painful experiences involved what we might call *social pain*—pain of a loved one's dying, of being dumped by someone you loved, or of experiencing some kind of public humiliation in front of others. Why do we associate such events with the word *pain*? When human beings experience threats or damage to their social bonds, the brain responds in much the same way it responds to physical pain.

Birthing Big Brains

Why are our brains built in such a way that a broken heart can feel as painful as a broken leg? One reason why being rejected hurts so much is that the larger brain was the easiest way for evolution to make us smarter. Having a larger brain, relative to one's body size, is instrumental to one species' being smarter than another species. And as we discussed, adult humans have a particularly large brain relative to their body size. Giving birth to a baby with a big brain is not easy, as any woman who has given birth can attest. The rest of the body passes through the birth canal "relatively" easily, but the head can often barely make it out. Given the shape of the female pelvis, infants have to be born when they are because if the brain were to keep growing, human infants would not be able to be delivered.

The human infant brain is typically only a quarter of its adult size. That means the great majority of the brain's development happens after we are born. It matures as much as is possible in the womb, but this still leaves the lion's share of developmental work to be done after birth. The upside to this state of affairs is that our brains are finished being built while they are immersed in a particular culture, allowing our brains to be fine-tuned to operate in that specific environment. The downside to an immature brain is that babies are ill equipped to survive on their own. Human babies are born completely helpless and stay that way for years. In fact, we have by far the longest period of immaturity of any mammalian species. (Many parents will be happy to tell you that this period of immaturity lasts well into the twenties!) And it is true that the human prefrontal cortex does not finish developing until the third decade of life. While humans are the mammals born the most immature, all mammals share this characteristic to some degree. Our tendency to be born with immature neural machinery extends back 250 million years to the very first mammals and this was the first step in making us the social creatures we are today.

Inverting Maslow

In 1943, Abraham Maslow, a famous New England psychologist, published a paper in a prestigious journal describing a *hierarchy of needs* in humans. The hierarchy he identified is typically depicted as layers of a pyramid (see Figure 3.1). Maslow suggested that we work

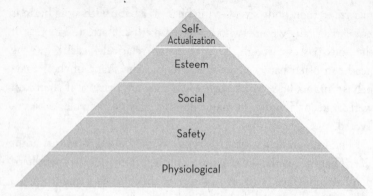

Figure 3.1 Maslow's Hierarchy of Needs. Adapted from Maslow, A. H. (1943).
A theory of human motivation. *Psychological Review,* 50(4), 370.

our way up the pyramid of needs, satisfying the most basic needs
first and then, when those are satisfied, moving up to the next set
of needs.

At the bottom of the pyramid are physiological needs like
food, water, and sleep. The next level of the pyramid focuses on
our safety needs, such as physical shelter and bodily health. Physiological and safety needs are really fundamental needs with a
capital *N.* No one can do without them. The rest of the pyramid
consists of "nice if you can get them" needs, or needs with a small
n. My son may say he needs another scoop of ice cream, but really he just wants one; he will survive without it (even if he thinks
he won't). In Maslow's pyramid, the remaining needs—the extra
scoops of ice cream—are love, a sense of belonging, and being
esteemed. *Self-actualization* (that is, reaching one's full potential)
is the cherry on top.

Ask people what they need to survive, and there is a very strong
probability that they will produce answers from the bottom tiers of
the pyramid, like food, water, and shelter. Infants need food, water,
and shelter too. The difference is that infants have no way of get-

ting these things for themselves. They are absolutely useless when it comes to surviving on their own.

What all mammalian infants, from tree shrews to human babies, really need from the moment of birth is a caregiver who is committed to making sure that the infant's biological needs are met. If this is true, then Maslow had it wrong. To get it right, we have to move social needs to the bottom of his pyramid. Food, water, and shelter are *not* the most basic needs for an infant. Instead, being socially connected and cared for is paramount. Without social support, infants will never survive to become adults who can provide for themselves. Being socially connected is a need with a capital *N*. Like the default network in Chapter 2, this restructuring of Maslow's pyramid tells us something critical about "who we are." Love and belonging might seem like a convenience we can live without, but our biology is built to thirst for connection because it is linked to our most basic survival needs. As we will see, connection is the first of three adaptations that support our sophisticated sociality, but our need for connection is the bedrock upon which the others are built.

Pain

A doctor has three patients waiting to see her. The first patient comes in complaining of a headache. The doctor says, "Take two Tylenol, and call me in the morning." The second patient hobbles in favoring one leg and says, "Doc, I think I sprained my ankle. What should I do?" The doctor says, "Take two Tylenol every day, and call me in a week." The third patient walks in having difficulty maintaining her composure, and she says, "Doc, I've got a broken heart. What should I do?" Without missing a beat, the doctor says, "Take two Tylenol every day, and call me in a month." True story? Of course not. No doctor would prescribe a painkiller to deal with feelings of rejection. But the story is instructive because our reactions to it reveal our intuitive theory of pain.

Pain is a fascinating phenomenon. On the one hand, it is very unpleasant, sometimes excruciatingly so. Yet it is one of the most fundamental adaptations promoting our survival. Nearly 20 percent of adults live with chronic pain, leading to countless lost workdays and deep depressions. One recent study estimated that pain was responsible for over $60 billion in lost productivity per year within the United States alone. As awful as chronic pain can be, not feeling pain is far more disastrous. Children born with congenital insensitivity to pain are incapable of feeling pain and often die in the first few years of life because they injure themselves relentlessly, often falling victim to deadly infections.

Pain is also at the center of many of society's moral decisions. Innovations in how the death penalty is carried out, from the guillotine to lethal injection, are considered progress because they minimize pain to those on death row. As a society, more of us are all right with sentencing someone to state-sponsored death than to sentencing someone to state-sponsored pain. Whether or not a fetus can feel pain has been brought into the debate about abortion. Similarly, which animals are able to feel pain is often invoked in discussions of which animals can be sacrificed for food.

But in each of the aforementioned cases, we're talking about physical pain. What is our reaction to *social pain*, the pain of real or perceived damage to our social connections? When someone says, "He broke my heart," we understand this as a metaphor. No one mistakes this for a medical emergency ("We've only got moments to repair this broken heart. Nurse, get me 200 volts to the paddles. Clear!"). Most of us believe social pain isn't real pain; here, "pain" is just a figure of speech.

Real pain (that is, physical pain) serves an essential role in our survival. For every need with a capital N, there is a corresponding physical pain with a capital P that we feel when the need is not being met. A lack of food leads to hunger, and this painful state of deprivation motivates us to find food. A lack of water leads to thirst, which when unquenched can be similarly painful and motivating.

Physical injury leads to bodily pain, which motivates us to find shelter and rest so that our body can heal itself.

If our social needs really are basic survival needs with a capital *N*, then unmet social needs should be experienced as a pain with a capital *P* too. This sentiment was expressed by noted neuroscientist Paul MacLean, who wrote, "A sense of separation is a condition that makes being a mammal so painful." Is there a link between the pain of physical harm and the pain of social harm?

Is Social Pain Real?

I have been studying social pain with my wife, psychologist Naomi Eisenberger, for the better part of a decade now (she more than I). In the next several pages, I will try to convince you that social pain is a kind of real pain. But I have to be honest with you—to this day, part of me still finds this hard to accept. Physical and social pain seem as if they are worlds apart. Every time I experience a physical pain, I can point to a place on my body where I am feeling the pain; presumably there is some kind of disturbance or tissue damage at the spot where the pain is coming from. When I feel social pain, where should I point to?

In reality, physical pain is no more physical than any other psychological experience we have, such as seeing a red square, discovering the serenity of meditation, or anticipating a great first date. There are two distinct but equally important ways to interpret the preceding statement. First, pain is *less* physical than we typically assume. We know this because pain can be dramatically modified through the power of suggestion, via hypnosis or placebo treatments. There have actually been surgeries performed on individuals under the influence of hypnosis only, with no anesthesia and no pain. In pain experiments, simply expecting that a shock you are about to receive will be very painful can make the shock feel more painful than it would be otherwise. Various psychological disorders such as anxiety and depression commonly alter how sensitive we are to physical pain. Pain may not be *all* in your mind, but it is a lot more in your mind than most people realize.

There is also a second interpretation of the statement equating the physicality of pains and anticipating a first date. What we take as purely psychological events are *more* physical than we typically assume, in the sense that all psychological events are rooted in the physical processes of the brain. The serenity of meditation is the result of biochemical and neurocognitive processes occurring in the brain and body. If the joy of connecting with others did not have a physical basis in the brain, then there would be no way for a pill to shape and induce those feelings, and yet that is exactly what the drug ecstasy does. And how else can we explain that a drink that selectively depletes the brain's serotonin will render a person more sensitive to insults moments later? I don't mean to suggest that the psychological aspects are somehow done away with. I am not a reductionist. Rather, in daily life we tend to create an artificial separation between things like pain and emotion. Pain, emotion, and all that we experience are necessarily simultaneous expressions of psychological and physical processes.

Starting from this view, it is not beyond the realm of possibility that something as seemingly abstract as social pain could be just as tangible and just as painful, from the brain's perspective, as physical pain. I do not mean to suggest that physical and social pain are identical. No one has ever broken his arm and confused that with having been dumped by his girlfriend. And memories of social pain are much more intense than memories of physical pain. Different kinds of pain feel differently and have distinctive characteristics. What I am suggesting is that social pain is real pain just as physical pain is real pain. Understanding this has important consequences for how we think about the social distress that we and those around us experience.

One of the obvious hints that social pain is similar to physical pain is the language we use to talk about social pain. Most of the words we use to describe feelings of social rejection or loss involve the language of physical pain. We say, "She broke my heart," or "He hurt my feelings," or that a girlfriend's leaving "was like being punched in the gut." Psychologists are discovering that language that sounds metaphorical is often less metaphorical than first supposed. When it comes to social pain, the language of physical pain is the metaphor du jour all around the world. This is true in Romance languages like Spanish and Italian, which share roots with English, as well as in Armenian, Mandarin, and Tibetan. It is unlikely that this metaphor would spring up again and again across the globe if there were no connection.

Wire and Cloth

A second piece of evidence that social pain is real pain is the separation distress that mammalian infants show when separated from their primary caregivers. Anyone who has had a baby has observed the intense and relentless crying and distress that can occur when a mother leaves her child. In the 1950s, psychologist John Bowlby developed the concept of *attachment* to explain the observations he and others made during World War II of orphans and abandoned children living in residential nurseries, who did not receive the warmth, love, and affection that children typically experience. He posited that each of us is born with an attachment system responsible for monitoring our proximity to a caregiver and that this attachment system sounds an alarm when we lose that proximity. Internally, that alarm manifests itself as painful distress, which quickly becomes loud crying, a separation distress call that serves to alert the caregiver to retrieve the infant.

Attachment distress is distinctly social; it is as much a signal for those around the infant as it is for the infant itself. And like a walkie-talkie, the attachment system works only if child and caregiver stay connected. If babies were born with attachment systems that faded away in adulthood, then the cries of babies might fall on emotionally deaf ears. Fortunately, the same attachment system that causes us to cry as infants when we are separated from our caregiver also causes us to respond to our own baby's cries once we are grown. We all inherited an attachment system that lasts a lifetime, which means we never get past the pain of social rejection, just as we never get past the pain of hunger. We have an intense need for social connection throughout our entire lives. Staying connected to a caregiver is the number one goal of an infant. The price for our species' success at connecting to a caregiver is a lifelong need to be liked and loved, and all the social pains that we experience that go along with this need.

One of Bowlby's contemporaries, psychologist Harry Harlow, examined primate attachment processes in some of the most striking psychological studies ever conducted. He was working with rhesus monkeys in the 1950s when behaviorism was at its heyday, and concepts like love and attachment were taboo among animal researchers. An infant's apparent emotional attachment to its mother was understood as associative learning. In other words, the warmth, smell, and feel of the mother were thought to be associatively linked with primary reinforcers like food. By this account, infants "care" about their mothers only because of the statistical association between the presence of mom and the satisfying of their needs. By this view, if a poster of Barry Manilow were present whenever feeding took place, infants would become Barry Manilow fans because they would associate him with being fed. Harlow didn't buy this account, so he put it to a test.

Newborn monkeys were raised apart from their mothers. Substituting for the mother were two surrogate "monkeys" that Harlow built in his lab. One surrogate was a wire-mesh frame that was constructed roughly in the shape of an adult monkey and that provided the milk the newborns needed for survival. The other surrogate was a wooden block that was covered in a layer of sponge rubber, with an outer surface of terry cloth, also roughly in the shape of an adult monkey. This cloth mother provided no milk. Harlow then kept track of which surrogate the infants became more attached to: the one associated with nourishment or the one that felt a little more like a real mother monkey. The results were unambiguous and profound. Soon after birth, infant monkeys were spending nearly 18 hours a day in contact with the cloth monkey and almost no time at all with the wire monkey that provided food. The food association theory of why infants cling to mothers was clearly wrong. These monkeys were attached to the thing that felt most like a real monkey, regardless of the sustenance it provided.

Since Harlow's work, social attachment has been identified in a variety of mammalian species. Given that all mammals are born incapable of caring for themselves, they all have a similar need to stay connected with a parent or caregiver. Across a wide range of mammalian species, including rats, prairie voles, guinea pigs, cattle, sheep, nonhuman primates, and humans, scientists have discovered *separation distress vocalizations*—cries made when infants are separated from caregivers, typically leading to retrieval by the caregiver. Separation also leads to increased production of *cortisol* (a stress hormone) and long-term social and cognitive deficits. Children under the age of five who are separated from their parents by lengthy hospital stays can develop long-term behavioral and literacy deficits. And children who lose a parent show elevated cortisol responses a decade later. This type of early childhood stressor can also lead to brain alterations in a key region related to self-regulation in social contexts, as I'll discuss in more detail in Chapter 9.

In 1978, Jaak Panksepp, a luminary in the field of affective neuroscience, hypothesized that social attachments functioned by piggybacking onto the physical pain system and did so through opioid processes. Opioids are the brain's natural painkillers. Their production and release diminish the experience of pain. This is why morphine, a synthetic opiate, is such a powerful painkiller. Like all opiates, morphine is powerfully addictive. Panksepp noted the parallels with attachment processes in animals. Separation appears to cause drug withdrawal–like pain, whereas reconnection appears to act like a painkiller. Additionally, infants and caregivers show a reciprocal devotion that fits the description of addiction.

Panksepp first tested his social pain hypothesis in a group of puppies. When the puppies were socially isolated, they produced separation distress cries. However, when the puppies were given low doses of morphine, the separation distress cries were largely eliminated. Since then, nonsedating levels of opiates have been shown to reduce separation distress cries in a variety of mammalian species. Moreover, reconnection between mother and infant increases opiate levels, naturally, in both parties. This suggests that the same neurochemical that is instrumental in alleviating the distress of physical pain may also be central in alleviating the distress of social separation in infants. This was the first hard evidence that social and physical pain are treated in similar ways by the brain.

The Anterior Cingulate Cortex and Human Pain

When we think about social pain in humans, there is a common montage of cinematic moments that easily come to mind. We think about being picked last for a team in gym class, being dumped by a significant other, or the death of a loved one.

For obvious reasons, we do not conduct experiments with humans that involve giving people morphine after they have been rejected, excluded, or cheated on. Instead of manipulating opioid levels artificially, as Panksepp had done with puppies, Naomi Eisenberger and I turned to fMRI to study how the experience of social pain is represented in the human brain.

In trying to understand the links between social and physical pain, we have focused primarily on a brain region called the *dorsal anterior cingulate cortex,* or dACC (*dorsal* means toward the top of the brain, and *anterior* means toward the front of the brain), and to a lesser extent on the *anterior insula,* or AI (see Figure 3.2). The *cingulate cortex* is a long brain structure that stretches from the back to the front of the brain, hugging the *corpus callosum* on the *midline,* or middle, of the brain. The word *cingulate* comes from the Latin word *cingere,* which means belt or girdle, and the cingulate looks like a belt for the corpus callosum. To get a better sense of these regions, try searching for them with Google Images, and scan through the images that come up. These pictures can help you visualize where the brain regions are in relation to one another, beyond what I can show here in a single figure. There are literally countless pictures of every brain region available on the Internet.

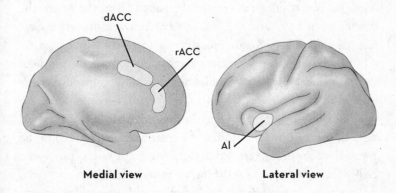

Figure 3.2 The Dorsal Anterior Cingulate Cortex (dACC), the Rostral Anterior Cingulate Cortex (rACC), and the Anterior Insula (AI)

There are four reasons why an investigation of the links between social and physical pain would lead to the ACC (anterior cingulate cortex) in general and to the dACC in particular. First, the ACC is one of the neural adaptations that distinguishes mammals from our reptilian ancestors. We have cingulates and reptiles do not. It makes sense to expect that new psychological processes first emerging in mammals, like attachment and social pain, might be linked to new mammalian brain structures like the ACC. Second, the ACC has the highest density of opioid receptors of any region in the brain, so it makes sense that physical and social pain may well be linked to this specific region. Third, it has been shown that the dACC plays an instrumental role in the experience of physical pain. Last, the dACC has been linked to mother-infant attachment behavior in various nonhuman mammals. Let's take these last two roles of the dACC in turn.

Over the last two decades, a great deal has been learned about the neuroanatomy of pain processes in the human brain. There are separate sets of cortical brain regions involved in the sensory and distressing aspects of pain. The *sensory aspects of pain* tell us where in the body the pain is coming from and how intense the stimulation is. Two regions residing in the back half of the brain, the *somatosensory cortex* and the *posterior insula*, track the sensory aspects of pain. The somatosensory cortex maps the different parts of our body, with distinct regions responding to pain in your legs, hands, or face. (The same regions also respond to nonpainful touch to the corresponding areas.) The posterior insula keeps track of pain sensations in our internal organs and viscera (that is, our gut feelings). In contrast, the dACC and the anterior insula, located in the front half of the brain, respond to the *distressing aspects of pain*—the feeling that makes pain something we really don't like.

Because pain feels like a single feeling while we are experiencing it, it's counterintuitive to imagine that there really are separable components to our experience of pain. This is a general trick of how the brain works. There are commonly multiple distinct components to any experience, but by the time it reaches consciousness, it is integrated into something that feels like one coherent event.

Imagine watching a person cross the street. It feels like a single fluid perception. In fact, many different brain regions are working together to orchestrate this experience. Some regions of the visual cortex code for all the lines and edges you see (vertical, horizontal, and diagonal lines). Another region keeps track of the color elements. Yet another takes in the motion from the scene you are watching. And each of these components can be knocked out while still leaving the others intact. We know this because of rare neuropsychological case studies that involve damage to a circumscribed brain region. For instance, there are patients with damage to motion perception centers who experience the world as a series of still photos, full of color and detail but with no intervening motion.

Similarly, neuropsychological cases have helped us figure out the distinct contributions of the dACC and the somatosensory cortex when it comes to pain. In the 1950s, neurosurgeons began performing a procedure called a *cingulotomy* on some patients with intractable pain. In this surgery, part of the dACC is removed or disconnected from surrounding areas. This surgery has been successfully used to treat depression and anxiety. But its greatest utility has come for individuals with chronic pain conditions that were not amenable to other kinds of treatment. The most striking thing about cingulotomies is the experience that chronic pain sufferers have postoperatively. They report that they still feel pain, and they can point to where it is on their bodies and indicate how intense it is. But also they report that the pain now is "not distressing," "not particularly bothersome," and "doesn't worry me anymore." For anyone with an intact dACC, it's nearly inconceivable that an individual could feel pain without experiencing the pain as distressing or bothersome, but that seems to be exactly what a cingulotomy allows. If removing or disconnecting a dACC selectively removes the distressing component of pain, this outcome implies that an intact dACC is central to this distress.

In another case, a stroke victim with selective damage to the somatosensory cortex on the right side of his brain (which keeps track of the left side of one's body) experienced pain-related changes that were the reverse of those associated with cingulotomies. As painful stimulation was applied to his left arm, he reported that he was receiving a "clearly unpleasant" feeling from somewhere between his fingertips and his shoulders. But he was unable to give a more precise location. And when asked to characterize the nature of the pain—hot, cold, or pinprick-like—he could not choose any of these. He was distressed by the pain, but he didn't know where it was on his body or how to describe it. If we were to make an analogy to reading a book, the somatosensory cortex seems responsible for understanding the type of story we are reading (thriller, detective novel, sci-fi) and its content, whereas the dACC is more responsible for one's emotional reaction to the narrative. We know that these reactions are separable since we can remember our emotional reactions to books long after we forget the plotlines.

The Anterior Cingulate Cortex and Attachment

The dACC and the ACC more generally are also critically important to attachment-related behavior for both mothers and their young. As we discussed earlier, mammalian young produce distress vocalizations when they are separated from their mothers or caregivers. Reptiles, from which mammals evolved, do not produce distress vocalizations, or any vocalizations at all—they are mute. And it's a good thing because most reptilian parents would likely eat their young if the young reptiles drew attention to themselves. The fact that mammalian crying serves as a cue for maternal support, rather than as a dinner bell, is a major evolutionary difference.

Neuroscientist Paul MacLean experimented with the effects of *lesioning* (that is, surgically disconnecting) different parts of the *medial frontal cortex* (which includes the ACC) on the distress vocalizations produced by squirrel monkeys when socially isolated. The only region whose removal consistently eliminated distress calls was the dACC. When other regions were lesioned while leaving the dACC intact, the distress calls continued. MacLean noted that all of the monkeys continued to produce other kinds of vocalizations ("yaps," "cackles," and "shrieks") postoperatively, regardless of which region he lesioned, indicating that these regions were not, per se, involved in the physical production of vocal sounds.

If removing the dACC eliminates distress calls, then one would think the electrical stimulation of the same region would generate them. And this is exactly what happens. When the dACC is stimulated in rhesus monkeys, they elicit the *köö*, a call that is specific to social isolation. In contrast, a warning call—a different kind of call—was elicited by stimulating other brain regions, but never by stimulating the dACC.

From these studies, we can begin to see the potential consequences to the infant's ability to form and maintain attachment bonds if the dACC is damaged. Isolated infants who don't cry are at a much greater risk of being left behind. And if the mother's dACC is damaged, she is less able to receive the infant's call on her end of the attachment walkie-talkie. To examine the effects of parents' dACC lesions on infants, female rats in one study were treated in one of three ways prior to giving birth. Some received cingulate lesions, some received noncingulate lesions (that is, lesions to other brain regions), and a third group received no surgery at all. The focus of the study was to examine how these lesions would affect the survival rates of the new pups born to these different types of mothers. The experimenters increased the harshness of the environmental conditions by adding heat and wind elements in certain parts of the cage to simulate conditions that might exist outside the lab.

Nearly every one of the pups of the mother rats who had not undergone surgery survived the first week. When the heat blasts hit their part of the cage, these mothers would corral all their pups over to an unaffected part of the cage. Moms with noncingulate lesions did almost as well, although some of their pups did not make it. But the consequences of the cingulate lesions were devastating. In this condition, only 20 percent of the pups survived the first two days after birth. These moms would not nurse their young, they built poor nests, they did not collect their pups when they strayed from the nest, and they dealt poorly with protecting their young from heat and wind. These mothers were unresponsive to the needs of their young. The difference between life and death for the pups was literally determined by whether their mothers had an intact cingulate or not. As an aside, if you find yourself distressed by this story, it probably means your own dACC is intact.

Cyberball

As suggestive as this animal data is, it does not tell us whether social pain is linked to the experience of physical pain in humans. Around 2001, Naomi Eisenberger and I decided to try to answer this question. We had just received a grant to study the role of the ACC in social cognition. We knew we wanted to study social rejection, but we had not come up with an ideal way to study it while someone was lying inside an MRI scanner.

As is often the case in science, random events intervened and changed the course of our research. We were at a conference in Australia that neither one of us really belonged at. It was there that we heard Kip Williams talk about a new experimental paradigm he had created for studying social rejection. It was entirely Internet based, yet it was highly effective at producing feelings of social rejection, so it translated well into the fMRI scanning environment.

Kip Williams's paradigm was called *Cyberball,* and it was a variant of a behavioral paradigm he had already been using successfully. In his first studies, a subject would show up and be told to wait for a few minutes. In the waiting room, two other people were already sitting, waiting for the same study. In reality, the other two people were what psychologists call *confederates*, which means they were pretending to be subjects and were actually working for the experimenter. One of the confederates would appear to "spontaneously" discover a tennis ball and would throw it to the other confederate, who would then toss it to the actual participant. Over the next minute or two, the three of them would toss the ball around in a triangle. However, at a prearranged time, the two confederates would stop throwing the ball to the real participant, and instead they would throw it back and forth to each other.

Imagine you are the person who has been left out of the game that you were all playing so nicely. On the one hand, you might think, "Who cares? It's not a real game and I don't know these people—they are complete strangers." That would be a very rational response, and undoubtedly some participants try to rationalize their sudden exclusion in this way. Yet, based on the measures Williams took, it was clear that these outcasts were in fact feeling social pain. It hurts to be left out, even in such a trivial way. After running a few of these waiting room studies, Williams created *Cyberball*, which replicated this scenario digitally. When playing *Cyberball*, the participant believes she is throwing the digital "ball" around with two other real people connected over the Internet. But in actuality she is playing only with preprogrammed avatars (see Figure 3.3) that stop throwing her the ball after a short while.

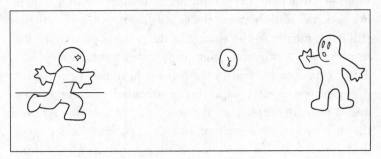

Figure 3.3 *Cyberball*

We had people play *Cyberball* while they were inside an fMRI scanner. The subjects believed that they and two other individuals were simultaneously having their brains scanned while they played the videogame over the Internet. We told them we were interested in how brains coordinate with one another to perform even simple tasks like ball tossing. The individuals had no idea they were about to get rejected in the scanner. But after a few minutes of throwing the ball around, the other "players" stopped throwing the ball to the actual participant.

After participants were rejected, they got out of the scanner, and they were taken to a room to answer questions about their experience. Frequently, these individuals would spontaneously start talking to us about what had just happened to them. They were genuinely angry or sad about what they had gone through. This was unusual for an fMRI study back then because most tasks didn't generate personal emotional reactions. We had to pretend that we hadn't been paying attention to what had happened in the scanner because we did not want their answers to the questions they were about to be asked to be contaminated by anything we might say.

We spent the better part of the next year analyzing the data, but there was a single moment when we knew we might be on to something interesting. Naomi and I were in the lab late one night, and my graduate student Johanna Jarcho was analyzing her data from a physical pain study on the next computer. We were all looking back and forth between the two data sets when we noticed a striking similarity in the results. In the physical pain study, participants who experienced more pain distress activated the dACC more. The same was true in the social pain study, as participants who experienced more social distress when rejected activated the dACC more. In the physical pain study, participants who activated the right ventrolateral prefrontal cortex experienced less physical pain. Similarly, in the social pain study, participants who activated the right ventrolateral prefrontal cortex experienced less social pain. Finally, in both studies, participants who activated the prefrontal region more activated the dACC less.

Both studies were telling us the same thing. When you experience more pain, there is more activity in the dACC. Lots of studies had shown this before ours—but ours was the first study to show that this was true not only for physical pain but for social pain as well. In both cases, a person's ability to regulate the distressing aspects of pain was associated with increased ventrolateral prefrontal activity, which in turn seem to mute the dACC response. Looking at the screens, side by side, without knowing which was an analysis of physical pain and which was an analysis of social pain, you wouldn't have been able to tell the difference.

These findings highlighted one of the things fMRI research can do to help us understand the human mind in general. It can illuminate when two mental processes that seem different actually rely on common neural mechanisms, suggesting they are more psychologically intertwined than we would have guessed. Here, the mammalian need to recognize social threats appears to have hijacked the physical pain system to do what the pain system always does—remind us when there is a threat to one of our basic needs.

What Does the dACC Really Do?

When our *Cyberball* paper was published, it propelled our careers. Newspapers and television shows wanted to interview us. A number of documentaries being made about pain or social connection wanted to include a segment about our work. We even got invited back to the conference in Australia that had inspired the study so that Naomi could present it.

Nevertheless, lots of scientists didn't buy our findings that the dACC supported the experience of social pain or that social and physical pain shared underlying processes. It's natural for scientists to be skeptical of a finding before it is replicated. But in our case, the skepticism was less about waiting for replications and more about not believing the story was plausible. At the time, the dominant theory of dACC function implied that it had little to do with pain processing, social or physical. This account largely ignored all of the cingulotomy and animal work from the 1950s as if there were a statute of limitations on the validity of those scientific findings.

In the mid- to late 1990s, several neuroimaging studies were published suggesting that the dACC performed two closely related cognitive functions: *conflict monitoring* and *error detection*. Here's a simple demonstration. Say the following words out loud: *now, how, cow, wow, mow*. If you hesitated when you got to *mow* but pronounced it correctly, that's conflict monitoring (that is, you detected that there was a conflict between your impulse and the correct response). If you pronounced it incorrectly and then said, "Oops, that's not a word—it was the leader of the Chinese Communist revolution," you just engaged in error detection.

In 2000, a scientist named George Bush (no relation to the former presidents) published a seminal paper on the function of the dACC. Citing many neuroimaging studies of cognitive control, he too concluded that the dACC plays a key role in cognitive processes like conflict monitoring and error detection. It was a conclusion that still holds up very well a decade later.

Bush's review also concluded that the dACC does *not* play a role in emotional processes. Processes related to emotion were identified with another part of the cingulate: the rostral anterior cingulate cortex (rACC). Superficially, this seemed like a parsimonious division of labor. Psychologists have long enjoyed dichotomizing processes into cognitive and emotional variants (such as thinking versus feeling), as if these were mutually exclusive phenomena. Bush drew this conclusion from several studies that appeared to show emotional processing localized to the rACC, but not to the dACC. But that conclusion doesn't hold up even based on the data that was available back then. All but three of the emotion studies reviewed focused on psychiatric populations, who may not be representative of how healthy brains respond. The majority of the nonpsychiatric studies included actually showed that the dACC *was* involved in affective processes. Moreover, several other neuroimaging papers of emotion or pain distress that were left out of the review, but that had already been published at the time, clearly pointed to dACC involvement in emotion or pain distress. As preferable as it might have been to link the dACC with cognition and the rACC with emotion, the truth is more complex.

Our Alarm System

A year after our first paper on social pain came out, Naomi and I published a paper on a new model of dACC function that sought to characterize both the affective and the cognitive functions of this region. We characterized the dACC as an *alarm system*.

Let me tell you about a few of the lousy alarms in our house to illustrate what is necessary for a good one. We live in an older home with some quirks that still have not been fixed since we moved in a few years ago. First, we have a doorbell on our front door that doesn't work. If you stand close to the front door, you can hear a whisper-level sound of a circuit connecting when someone pushes the button, but that is all you will hear. Until the pizza delivery guy realizes he should try the metal knocker on the door, he just waits, assuming we can hear the doorbell when, in fact, we can't. I know we should get it fixed, but everyone figures out to use the metal knocker, so we have never been particularly motivated to do so. We also have a smoke detector that goes off every once in a while even when there is no smoke. This is especially annoying when the every once in a while is at 3 a.m.

These are both terrible alarms; each is missing one of the two vital components of a functioning alarm mechanism. An alarm needs a *detection system* that keeps track of whether some condition has been met or not. Smoke detectors commonly use a photoelectric detector that consists of an unbroken beam of light hitting a pho-tocell. When a sufficient number of smoke particles break the light beam, smoke has been detected. Given that our smoke alarm goes off at random times when there is no smoke in the house, there is something wrong with its detection system. An alarm also needs a *sounding mechanism* that is triggered by the detection system. The sounding mechanism in our smoke alarm works fine, obviously. But the sounding mechanism in our doorbell doesn't function, so we don't know when someone is at the door.

In our neural alarm system model, we proposed that the dACC is an alarm system that serves both to detect a problem *and* to sound an alarm. The smoke alarm needs to let everyone in the vicinity know that there might be a fire, to call 911, or just to flip the burgers so they stop burning. It has to be able to interrupt whatever else you are doing or focusing on. This is precisely what emotions do for each of us. The conscious distress of physical pain motivates us to take our hand off the stove; the pain of social exclusion motivates us to work to reconnect with others.

Detecting conflicts and errors is often a source of emotional experience. Getting a B on a test isn't intrinsically emotional, but if you expected to get an A+, it will most likely cause distress. It occurred to us that the conflict monitoring and error detection studies that pointed to the dACC's role in cognition might have also produced emotional responses, but perhaps the studies overlooked them because these emotional responses were never measured. So we decided to measure them.

Bob Spunt, then a graduate student in our lab, ran an fMRI study with Naomi and me in which he used a conflict monitoring/error detection procedure called the *stop-signal task*. (This task is a variant of the *go/no-go task* described in Chapter 9.) On most trials the task was incredibly simple. An arrow appeared on a computer screen pointing to the left or right; when it did, a corresponding key on the keyboard had to be hit as quickly as possible (one key for left, one key for right). These trials went by at a rapid clip, about one per second, and they were easy. A quarter of the trials, however, required a different response and were trickier. On these, a *stop-signal tone* was played after the arrow appeared. This tone indicated that participants should ignore the arrow and not press any button on that trial. It was a signal indicating that the participants should stop, just for that one arrow. This is akin to a traffic light turning yellow just as you are getting to the intersection; the changed light indicates that you need to override the plan you have already set in motion. On early trials, the tone was played about 250 milliseconds after the arrow appeared. If this gave participants enough time to stop themselves from hitting an arrow key, the tone was shifted so that it came later. The task kept changing until the tone came long enough after the arrow key that participants couldn't help but mistakenly hit the arrow key when they shouldn't have, about half of the time. Participants couldn't win. The better they were at this, the harder the task became. Personally, I find the task absolutely maddening, which is why it was perfect for our purposes.

After every 16 trials that included 4 of the dreaded stop trials, participants were asked to what extent the just-completed block of trials had made them feel anxious and frustrated. There were also *go-only blocks* that included no stop trials, and participants were always informed which kind of block was coming next. People knew whether the upcoming block was going to have the annoying stop trials or not.

In Bob's first analysis, he demonstrated that the *error trials* (that is, when people were meant to stop but failed to) produced a strong response in the dACC, just as countless prior studies had. Next, he used the frustration that participants expressed at the end of each block to see if there were brain regions whose activity was stronger during errors that were more frustrating, compared to errors experienced as less frustrating. Although the task didn't change much from block to block, people did report some blocks being more frustrating than others, and the activity in the dACC tracked this. The more frustrating the errors, the greater the dACC activity. No other region in the brain, besides the dACC, tracked the frustration participants experienced during the errors on this task. We also found some evidence suggesting that even on the other trials that did not require stopping, the dACC produced greater activity to the extent that participants were anxious. In other words, as participants became more anxious about the prospect of stop trials, we saw evidence of their anxiety in the dACC responses.

The results help us understand the functions of the dACC better. Historically, the dACC has been framed as supporting either cognitive or emotional functions, with recent trends supporting the former. We posited that the dACC supports both cognitive *and* emotional functions. Specifically, we suggested that the dACC is an alarm system with this region serving both as a detection system (cognitive) and as a sounding mechanism (emotional). The data here demonstrates that while the dACC is activated by a standard error detection task, the strength of activity in the dACC is also linked to the emotional experience of making an error.

Take Two Aspirin

Our basic findings linking social exclusion to dACC activity have been replicated in a number of studies and extended to people experiencing grief over the death of a loved one, remembering a recent romantic breakup, being negatively evaluated, and even just looking at disapproving faces. Toward the beginning of this chapter, I told the story of the doctor who had three patients, the first two with physical ailments and the third with a broken heart. The doctor prescribed painkillers for all three. In the context of a broken heart, this seemed farfetched. Nevertheless, when we give talks about our fMRI work on social pain, it is not uncommon for someone to come up after and open with some variant of "What do you tell someone who has just been rejected? Take two aspirin and call me in the morning?"

Although I was deliberately dismissive of this idea at the beginning of the chapter, the real answer is, "Well, yes, sort of." Nathan DeWall, together with Naomi Eisenberger and other social rejection researchers, conducted a series of studies to test out the idea that over-the-counter painkillers would reduce social pain, not just physical pain. In the first study, they looked at two groups of people. Half of them took 1,000 milligrams a day of acetaminophen (that is, Tylenol), and half of them took equivalently sized placebo pills with no active substances in them. Both groups took their pills every day for three weeks. Each night, the participants answered questions by e-mail regarding the amount of social pain they had felt that day. By the ninth day of the study, the Tylenol group was reporting feeling less social pain than the placebo group. Moreover, between the ninth day and the twenty-first day, the difference between the two groups kept widening. Neither group knew what they were ingesting. Yet taking the painkiller we reach for to make a headache go away seems to help make our feelings of heartache go away too.

This first behavioral study was followed by an fMRI study. Participants once again took either Tylenol or a placebo every day for three weeks and then were scanned while playing *Cyberball*. At first, they were included in the videogame for a few minutes, and then they were left out for the rest of the game. Those who had been taking placebo pills for three weeks responded similarly to the subjects in our earlier *Cyberball* fMRI studies. They showed greater activity in the dACC and the anterior insula regions of the brain when they were excluded from the game, compared with when they were included in the game. In contrast, those who had been taking Tylenol for three weeks showed no dACC or insula response to being rejected. Taking Tylenol had made the brain's pain network less sensitive to the pain of rejection.

Another study made the direct link between the dACC findings and Panksepp's original opioid hypothesis of social and physical pain. Naomi Eisenberger worked with Baldwin Way to find a genetic trait that relates to social pain. They focused on the mu-opioid receptor because of its role in medicating pain. Mice that have been bred to lack the mu-opioid receptor no longer respond to morphine. In humans, the experience of pain depends in part on the mu-opioid receptor gene (called OPRM1). Within this gene there are three variations ("polymorphisms") at a particular spot on the gene that alter how much the gene will be expressed. Each of us has two alleles that determine which polymorphism we have. We inherit one allele from our mother and one from our father. Each can be an A or a G; thus, each of us is an A/A, A/G, or G/G.

Prior pain studies have demonstrated that G/Gs are more sensitive to physical pain (for example, they require greater quantities of morphine to deal with postoperative pain).

Genetic samples were obtained from a group of individuals in order to determine which variant of the OPRM1 gene they had. They were also asked to indicate how sensitive they were to social rejection in their everyday life. Those with the G/G variant of the OPRM1 gene (that is, those likely to be more sensitive to physical pain) reported being more sensitive to social rejection than those with the other variants. A subset of these individuals also participated in an fMRI *Cyberball* study, and the same genetic pattern held with respect to the dACC and anterior insula activity they produced when rejected. G/Gs produced more activity in these regions when rejected than other participants. My sense is that the Tylenol and opioid studies were what really convinced a lot of scientists that social and physical pain are really making use of the same pain equipment in the brain. People may not know much about specific brain regions, but from personal experience most know something about painkillers. Tylenol's effects seem really selective to pain—it doesn't dull our mind or distract us from pain through pleasant feelings. It seems to zero in and target something specific to pain. To see these drugs diminishing our social pain as well as physical pain speaks strongly to the connection between the two kinds of pain.

Sticks and Stones

In the abstract, *Cyberball* seems like a trivial game with a trivial outcome. Two "strangers" that you have never met stop throwing a digital ball to you in the most boring game of catch you will ever play. How is this relevant to anything that matters in your life? Being included by others when playing *Cyberball* won't help you get better clothes, the job, or the girl. As a participant, you get paid the same for the study whether you are included or excluded. Everything about this study seems small and insignificant. But the implications are profound—that something so small produces such dramatic effects. Our sensitivity to social rejection is so central to our well-being that our brains treat it like a painful event, whether the instance of social rejection matters or not.

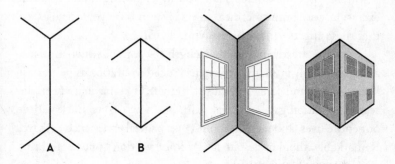

Figure 3.4 The Müller-Lyer Illusion

Consider visual illusions like the one in Figure 3.4, the Müller-Lyer illusion. People experience line *A* as longer than line *B* even though they are identical in length. Why? The human visual system makes various assumptions about what different visual cues in the environment imply, and it uses those assumptions to make sense of the complex world around us. In the Müller-Lyer illusion, the shape of the arrows at the ends of the lines is key. Line *B* has arrows that if extended, suggest that you are looking at the edge of two walls joining together close to you. In contrast, line *A* has arrows suggesting that two walls are meeting far away in the distance. Vertical lines of the same length hit your retina, but the arrows lead your brain to infer that line *A* is far away and line *B* is close up. The brain knows that identical retinal projections should be experienced as differing in size based on their distance. If our brains didn't do this, we would be terrified as people walked away from us and shrank until they disappeared.

Look at Figure 3.4 again. Now you know the trick, but the illusion persists. You will always see line *A* as longer than line *B*. This illusion is trivial, just like *Cyberball*, yet we continue to experience both effects. Kip Williams found that even when he told people they were just playing against a computer and that the computer was preset to reject them, people still experienced social pain. Making quick visual assessments and feeling pain in response to social exclusion were both so critical to survival in our evolutionary past that these effects cannot be easily mitigated.

We have already discussed at length the reasons why mammals, and particularly humans, need to feel social separation as painful. It keeps infants and caregivers close together. That may have been the reason evolution gave us social pain, but now we are stuck with it our entire lives, and it colors almost every social experience we have. Remarkably, though, despite its ubiquity, we don't understand this central aspect of our nature.

Imagine you have a thirteen-year-old son, Dennis, who is physically assaulted at school by a bully. The bully pushes Dennis down and hits him several times. What do you do when you find out? March into the principal's office? Call the police to press charges? Write to the local paper to express outrage at what is happening in our schools? Different parents would do any and all of these things. Now imagine that your Dennis is being bullied, but only in words. The bully never lays a hand on your son, but he teases him mercilessly, telling him that he is ugly and stupid and that no one likes him (none of these things are true). When Dennis reluctantly tells you about the teasing, what is your reaction then? Does it involve the police or local press? Not likely. More probably, your response will be something like this: "Just ignore him. You will be off to college in a few years, and he will probably be flipping burgers for the rest of his life." I don't mean to suggest that it isn't distressing to find out that your son has been teased, but it isn't the same as finding out that there was physical contact. We don't go to the principal, police, or press in this case because we don't think any of them will take action if it's just verbal teasing.

From a young age, we teach children to say, "Sticks and stones will break my bones, but names will never hurt me." But this isn't true. Bullying hurts so much not because one individual is rejecting us but because we tend to believe that the bully speaks for others— that if we are being singled out by the bully, then we are probably unliked and unwanted by most. Otherwise, why would all those others watch the bully tease us rather than stepping in to help support us? Absence of support is taken as a sign of mass rejection.

I bring up bullying because at a societal level, it is probably the most pervasive form of social rejection we have. Studies from around the world, including the United States, England, Germany, Finland, Japan, South Korea, and Chile, suggest that between the ages of twelve and sixteen, about 10 percent of students are bullied on a regular basis. Although bullying can involve physical aggression, more than 85 percent of bullying events do not. Instead, they involve belittling comments and making the victims the subjects of rumors. But victims of bullying suffer long after school is over and the bully has gone home. These individuals are seven times more likely than other children to report being depressed. They think about committing suicide more, and they are four times as likely as others to make a suicide attempt. Sadly, they are also more likely to succeed in their attempts. A 1989 Finnish study assessed the level of victimization among eight-year-olds from a sample of more than 5,000 students. Those who had been bullied at age eight were more than six times as likely to have actually taken their own lives by the age of twenty-five. Suicide-related thoughts are actually quite similar among those who have been victims of bullying and those who have been victims of chronic physical pain, further supporting the link between these two kinds of pain.

Throughout our lives, we are destined to experience different forms of social rejection and loss. Most of us go through multiple relationship breakups, and we typically spend a portion of those on the side of being left, rather than leaving. Such breakups often feel unbearable, and they can dramatically alter how we view ourselves and our lives for a long time after. Our Faustian evolutionary bargain allows us as humans to develop slowly outside the womb, to adapt to specific cultures and environments, and to grow the most encephalized brains on the planet. But it requires us to pay for it with the possibility of pain, real pain, every time we connect with another human being who has the power to leave us or withhold love. Evolution made its bet that suffering was an acceptable price to pay for all the rewards of being human.

CHAPTER 4

Fairness Tastes like Chocolate

Imagine you work for the law firm of Horn, Kaplan & Goldberg, and you are up for early promotion to partner. The promotion is likely to go to you or to Steve, a lawyer down the hall. You've got the numbers on your side. Your performance reviews have been stronger for six straight quarters, you have a better record in the courtroom, and over the past three years you have billed 30 percent more hours for the firm than he has. Steve has one thing going for him, an ace in the hole. He's Steve Goldberg, nephew of one of the senior partners in the firm. Steve's a good lawyer, and he deserves to make partner too, but you deserve it more. If there is only one slot, by all rights, it should be yours.

Not getting the position would mean missing out on a higher salary, but the social implications would make this outcome painful as well. Being passed over would hurt because it would feel like the firm's partners were rejecting you. Furthermore, this outcome would clearly constitute a social insult that everyone in the firm would know about. In this case, both your so-called basic needs and social needs would take a hit.

As it turns out, fortune shines down on you, and the keys to the executive washroom are handed to you instead of Steve Goldberg. The promotion comes with a large raise, and you and your husband will now have the money to move up to your dream home in the neighboring town with better schools for your children. No matter what your profession or professional aspirations, you have probably imagined or had this kind of moment.

Likely to be lost in your celebration over your newfound wealth and status is the role fairness might have played in your positive feelings after learning the outcome. The partners could have prioritized Steve's bloodline over your hard work and productivity. Even three-year-olds sharing cookies become upset when they are treated unfairly. Unfair treatment is demoralizing and often leads to a host of negative feelings. But does fair treatment produce positive feelings of its own? Fairness seems a bit like air—its absence is a lot more noticeable than its presence.

Being treated fairly is usually confounded with obtaining the better outcome, so it's hard to parse our positive feelings for each. If you are walking with a friend, and he picks up a $10 bill that you both saw lying on the ground and offers to share it with you, the bigger the cut he offers to you, the fairer it is. So when he gives you $5, how much of your happiness is due to getting $5 instead of $3 or $0 and how much is due to feeling valued by your friend? There have been a couple approaches to separating the joy of receiving more good stuff from the joy of being treated fairly. One approach has involved measuring the perceived fairness of events and the material benefit that people receive, separately. This has allowed researchers to use statistical analysis to see how both of these factors relate to people's feelings about the events.

In one study, participants were put on a team and independently performed an anagram task (for example, figuring out that LIOSAC can spell SOCIAL). The team was then paid based on their overall group performance. After receiving the money, the team members had to negotiate among themselves how to split their earnings. This was where things got tricky because some team members invariably scored higher than others and thus were more responsible for the payout the team received. Some team members found an equal distribution to be fair ("We all got the same amount"), and others found an equitable distribution to be fair ("We received according to our scores"). But whether the team members received a lot or a little, as long as they believed the process was fair, they had more positive emotions.

This same pattern has been observed in field research as well. Psychologist Tom Tyler found that defendants in court cases were happier with their courtroom experience if they believed they were treated fairly, even when the verdict did not go their way. How do we know these people really mean what they say? Maybe they don't really know how they feel, or maybe they are trying to give experimenters the answer they are looking for.

Golnaz Tabibnia and I thought that by scanning the brain, we would be able to garner additional evidence for or against the notion that fairness is intrinsically rewarding to us. We asked individuals lying in an MRI scanner to play an economic bargaining game that exposed them to both fair and unfair outcomes. They played a variant of the Ultimatum Game, in which two players have to agree how to split some amount of money, say, $10. One player, called the *proposer*, makes a recommendation about how much each of them should get, and the other player, called the *responder*, then decides whether to accept the offer. If the responder accepts, then both individuals get the amount suggested by the proposer. But if the responder rejects the offer, both players get nothing. A proposer might suggest that he get $9 and the responder get $1. And if you guessed that responders might be insulted by this kind of offer, you would be right. Responders commonly reject highly unfair offers, preferring to get nothing at all rather than let this insult go unpunished. This seems to fly in the face of rational self-interest, but it's what people do.

In our study, participants played the part of the responders and saw a series of offers from different proposers. We wanted to see if the brains of our participants would respond differently to fair and unfair offers. We faced the same challenge that we saw in our law firm example at the beginning of this chapter. An offer of $5 out of $10 is more fair than an offer of $1 out of $10, but the $5 offer is also much more lucrative. To deal with this complication, we varied the total amount to be split from offer to offer: responders would see offers of $5 out of $10, as well as offers like $5 out of $25. In both cases, the material amount of the offer was equivalent ($5), but the offers differed significantly in their fairness. By doing this, we could attribute neural differences to the effects of fairness rather than to the financial gain.

Most studies have used this paradigm to look at neural responses to unfair offers. Consistent with the social pain findings described in Chapter 3, these studies typically observe activity in the anterior insula and the dACC. However, when we looked at which regions were more active during fair offers as opposed to unfair offers, almost all of the regions we observed were part of the brain's reward network (see Figure 4.1). Being treated fairly turned on the brain's reward machinery regardless of whether it led to a little money or a lot.

An even more dramatic demonstration followed when a group of researchers from Cal Tech examined the neural responses of individuals as their own potential winnings were given to another participant. Ordinarily, this experience would be painful rather than pleasurable. Who wants to see one's own money taken away? As it happened, the individuals who won a $50 lottery at the beginning of the study and then saw the other participant not win any money showed robust activity in the brain's reward system when the lottery loser went on to win money on subsequent trials—even though it was at the participant's own expense. Subsequent wins by the lottery losers brought the losers' total earnings more in line with the participants' own earnings, and seeing a fair distribution of the winnings was more rewarding to participants than gaining more for themselves. In other words, fairness trumped selfishness.

Fairness is one of many cues that we have that we are socially connected. Fair treatment implies that others value us and that when there are resources to be shared in the future, we are likely to get our fair share. Fairness is clearly a more abstract sign of social connection than many others we could imagine, and it's important enough that our brain's reward system is sensitive to it. The same brain regions that are associated with loving the taste of chocolate or any other physical pleasures respond to being treated fairly as well. In a sense, then, fairness tastes like chocolate.

This chapter isn't about fairness per se, but rather about the various social signs, events, and behaviors that reinforce our connection to an individual or the group. Because these tend to activate the brain's reward system, they are referred to as *social rewards*. Just as social and physical pain share common neurocognitive processes, so to do physical and social rewards share common neurocognitive processes.

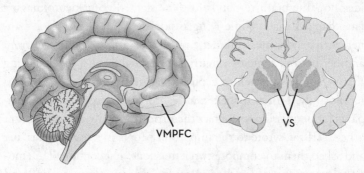

Figure 4.1 The Brain's Reward Circuitry
(VMPFC = ventromedial prefrontal cortex; VS = ventral striatum)

Oscar and Sally

In the 1984 film *Places in the Heart*, actress Sally Field portrayed a 1930s southern widow trying to keep her farm out of foreclosure. For her performance, she went on to win the Academy Award for best actress. Her acceptance speech was memorable for its enthusiastic earnestness. In the most famous line from Field's acceptance speech, she declared, "You like me. You really like me." Even if you didn't know who said it, I bet you have heard that line before and know it was uttered with a strong emphasis on the word *really*. It exemplifies the adulation that actors crave.

There are two errors in the previous paragraph, one more important than the other. The minor error: Sally Field did not actually say this line in her acceptance speech. The real line in her speech was, "I can't deny the fact that you like me, right now, you like me." We probably misremember the quote because of the other, more important error. It isn't just actors who are motivated by being liked—we all are. The misquote is so sticky because it exemplifies a central human need.

We all have a need to belong. Signs that others like, admire, and love us are central to our well-being. Until very recently, we had no idea how the brain responds to these signs. Recent neuroimaging has changed that. While lying inside the bore of an MRI scanner, perhaps the most dramatic positive sign that we can get from another person, short of a marriage proposal, is to read something that person has written to express their deep affection for us. In a recent study, Tristen Inagaki and Naomi Eisenberger asked participants for permission to contact their friends, family, and significant others. Tristen wrote to the important people in a participant's life and asked them to compose two letters: one that contained unemotional statements of fact (for example, "You have brown hair") and one that expressed their positive emotional feelings for the participant (for example, "You are the only person who has ever cared for me more than for yourself").

Subjects would then lie in the scanner while reading these letters written about them by several of the people they care about most. Our intuitive theories suggest there is something radically different about the kind of pleasure that comes from people saying nice things about us and the kind of pleasure that comes from eating a scoop of our favorite ice cream. The former is intangible, both literally and figuratively, while the latter floods our senses. Although there are surely differences between physical and verbal sweets, this fMRI study suggested that the brain's reward system seems to treat these experiences more similarly than we might expect. Being the object of such touching statements activates the ventral striatum in the same way that the other basic rewards in life do.

In a follow-up study, Elizabeth Castle and I looked at how rewarding these touching statements really were. We asked a group of individuals to bid money to try to win these statements. In the end, a large proportion of the participants were willing to give back their entire payment for the study, just to get to see these special words. We may give lip service to the power of money, but the power of knowing we are loved can be just as potent.

It is easy to imagine our reactions to getting this rarely shared positive feedback from the people who matter most to us, but would social feedback from complete strangers have the same effect? Surprisingly, yes. Imagine Penelope, a twelve-year-old, lying in a scanner watching as a series of faces of other kids appears on the screen. Penelope has never met any of the people she is seeing, but she is informed after seeing each face whether that person wanted to have an online chat with her. Participants like Penelope showed increased activity in the brain's reward system when finding out that those strangers wanted to have an online chat with them. These findings were remarkable for two reasons. First, the feedback was ostensibly from complete strangers who had seen the participant's picture and knew very little else about him or her. Second, the positive feedback led to reward activity even when the participants had no interest in having a chat with the other person. So even strangers we don't want to interact with activate the brain's reward system when they tell us they like us.

Others studies have suggested that our brains crave the positive evaluation of others almost to an embarrassing degree. Keise Izuma conducted a study in Japan in which participants in the scanner saw that strangers had characterized them as *sincere* or *dependable*. Having someone we have never met and have no expectation of meeting provide us with tepid praise doesn't seem like it would be rewarding. And yet it reliably activated the subjects' reward systems.

When participants in this study also completed a financial reward task, Izuma found that the social and financial rewards activated the same parts of the ventral striatum, a key component of the reward system to a similar degree.

Sally Field really was speaking for us all when she expressed her delight at being liked by others. Not only are we sensitive to the positive feedback of others but also our reward system in the brain responds to such feedback far more strongly than we might have guessed.

If positive social feedback is such a strong reinforcer, why don't we use it more often to motivate employees, students, and others? Why isn't it part of our employee compensation plans at work, for example? A kind word is worth as much to the brain in terms of rewards as a certain amount of money. So why isn't it part of the economy, like all other goods we assign a financial value to? The answer is that it isn't yet part of our theory of what people find rewarding. We don't understand the fundamentally social nature of our brains in general and the biological significance of social connection in particular. As a result, it's hard for us to conceptualize how positive social feedback will be reinforcing within the most primitive reward system of our brains.

When I was in graduate school at Harvard working in Dan Gilbert's lab, I remember Kevin Ochsner telling me that I didn't praise the younger students in lab enough. I remember thinking, "Who am I to praise or not to praise? I'm a fifth-year graduate student who has yet to publish a single paper. My praise is meaningless." Of course, Kevin was right. If a stranger saying we are "dependable" activates the reward system, imagine what praise from a boss, a parent, or even an unaccomplished slightly older graduate student will do. Of course, we all know that praise is a good thing, as long as it isn't too unconditional, but until very recently, we had no idea that praise taps into the same reinforcement system in the brain that enables cheese to help rats learn to solve mazes. And positive social regard is a renewable resource. Rather than having less of something after using it, when we let others know we value them, both parties have more.

Varieties of Reward

Though it might not seem so, money is a social reward—a reward for doing something of social value. Everyone who earns a salary is paid to do something that others want done, whether the person receiving the money is the biggest rock star in the world or her accountant. We all get paid to provide a service, not because we are doing what we want to do. Some of us are lucky enough to enjoy our work, but that is not why we get paid. When I was a new professor, I used to joke that if I were in charge, most academics wouldn't be paid at all because most of us enjoy it enough that we would do it for free. But we are paid because of the value of our work to others. Money is a social currency, just not an altruistic social currency.

Rewards can be divided into primary and secondary reinforcers. Things that satisfy our basic needs like food, water, and thermoregulation are known as *primary reinforcers*—they are an end in themselves, and the brain comes prepared to recognize these things as reinforcing without needing to be taught about them. When in a deprived state, all mammals work hard to obtain these primary rewards. *Secondary reinforcers* are things that are not initially rewarding in and of themselves but become reinforcing because they predict the presence or possibility of primary reinforcers.

If a rat is placed in a maze and has to choose whether to turn left or right in order find the cheese, it will do its best to learn where the cheese will be so it can up its odds of getting the reward each time. If the cheese is randomly placed during each trial, the rat cannot learn. But if the experimenter always places a little patch of red paint on the side of the maze leading to the cheese, the rat will learn to follow the signs. The red patch is not intrinsically rewarding, but if it consistently predicts where the cheese will be, the rat's reward system will start responding to the red patch.

Money is the world's most ubiquitous secondary reinforcer. You can't eat or drink it, and you would need an awful lot of it to keep you warm. Yet obtaining money is the best way adults can guarantee that their other basic needs will be met—it puts food on the table and a roof over their heads. Although money doesn't intrinsically satisfy any needs, it is often viewed as the most desirable reward of all. Perhaps getting money is like getting several rewards at once, because we can imagine countless ways to spend it.

So where does social regard fall within this typology of reward? It is probably both a primary and a secondary reinforcer. When your boss tells you how impressed he is with your work on the Davidson merger, it is easy to imagine this praise leading to a larger Christmas bonus. But studies like Izuma's suggest that social regard might be a primary reinforcer as well. The brain's reward system is activated as a result of such praise, even from strangers who have no control over that Christmas bonus. Evolution built us to desire and work to secure positive social regard. Why are we built this way? One possible explanation is that when humans, or other mammals, get together, work together, and care for one another, everyone wins. Given that other living creatures are the most complex and potentially dangerous things in our environment, a push from nature to connect with others in our species, an urge to please one another, increases our chances of reaping the benefits of group-based living.

Working Together

In the Pixar film *A Bug's Life*, an easygoing colony of ants is terrorized by a group of Mafioso grasshoppers demanding an unseemly cut of the colony's food in return for their "protection." Early in the movie, the eventual protagonist, an ant named Flik, stands up to the mob boss and is quickly put in his place; he is no match for the grasshoppers. The rest of the film focuses on the ants and other bugs learning to work collectively to defeat their tormentors. Predictably, after multiple failures, the ants succeed in ridding themselves of the grasshoppers by working together. While depicted anthropomorphically through the life of ants, it is a classic story of human courage and cooperation. When we pool our resources, we can do more together than we can alone. Cooperation is one of the things that makes humans special. Many species cooperate, but as Melis and Semmann write, no other species comes close "to the scale and range of [human] cooperative activities." Compared to the rest of the animal kingdom, humans are supercooperators.

Why do humans cooperate so often? Why do they cooperate at all? The easiest answer is that people cooperate when they stand to benefit directly from the cooperative effort. In *A Bug's Life*, the ants band together to defeat the grasshoppers, so that they will no longer have to give away their food. Similarly, two college students taking the same class may study for an exam together because each believes they will improve their test scores more with their combined effort than by studying alone.

There are other kinds of cooperative helping where the self-interested payoffs are less conspicuous. The *principle of reciprocity* is one of the strongest social norms we have. If someone does you a favor, you feel obligated to return the favor at some point, and with strangers we actually feel a bit anxious until we have repaid this debt. This is why car salesmen will always offer you a cup of coffee. By performing a small favor for you, they render you indebted to them, and the only thing you can really do for them in return is buy a car, yielding a commission worth far more than that cup of coffee. Obviously a free drink alone doesn't always lead to a purchased car, but it can nudge people in that direction. Similarly, we may cooperate with someone in such a way that, in the short run, we give up more than we gain, but with the expectation that we will benefit through reciprocity in the long run.

More interesting are the kinds of motivations that must be present when cooperating clearly reduces the benefit to oneself in the long run. Behavioral economists use a game called the *Prisoner's Dilemma* to illustrate this phenomenon. In this game, two players have to decide whether to cooperate with each other or not. How much money players earn depends on the combination of their decisions. Imagine there is $10 at stake for you and another player (see Figure 4.2). If you both choose to cooperate, you each get $5, and if you both choose not to cooperate, you each get $1. So far the decision to cooperate is easy. However, if one of you chooses to cooperate and the other chooses not to, the noncooperating defector gets the entire $10, and the cooperator gets nothing. In other words, if you choose to cooperate, there's a chance you'll look like a chump as the other person takes all the money.

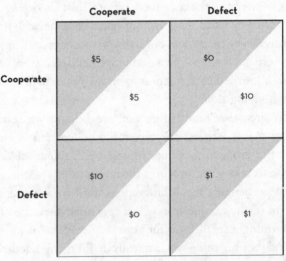

Figure 4.2 Prisoner's Dilemma Contingencies

Assume that you have never met the other player, do not get to discuss your decision with the other player, and will have no further interactions with that person after this one game. What do you do? If you want to make the most money and you assume the other person will cooperate, you should defect (because you'll earn $10 instead of $5). If you assume the other person will defect, then you should still defect (because you'll earn $1 instead of nothing). Regardless of what the other person does, you make more money by defecting. Nevertheless, multiple studies have shown that under these conditions, people still choose to cooperate more than a third of the time.

The Axiom of Self-Interest

How can we explain why folks cooperate, ensuring that they will earn less money and their partners will earn more? Are players who know the contingencies but still choose to cooperate irrational? If we believe nineteenth-century economist Francis Edgeworth's contention that "the first principle of economics is that every agent is actuated only by self-interest," then this cooperative behavior does seem irrational. And Edgeworth is hardly alone in putting selfishness front and center (and alone) as the basic motivation behind all of our actions. It's a common refrain. The eighteenth-century philosopher David Hume proposed that political systems should be based on the assumption that a man has "no other end, in all his actions, than his private self-interest." A century earlier, the philosopher Thomas Hobbes first formalized this account, charging that "every man is presumed to seek what is good for himself naturally, and what is just, . . . accidentally." This basic assumption is known as the *axiom of self-interest.*

A belief that self-interest guides everything we do leaves no room to explain the individuals who chose to cooperate, other than to suggest that they were irrational or they misunderstand the instructions. But how would this axiom explain the following findings? In some variants of the Prisoner's Dilemma, Player A is informed of Player B's decision before making his own. Not surprisingly, when Player A is told that Player B has chosen to defect, Player A always decides to defect (assuring himself of $1 instead of $0). What is surprising, though, is that when Player A is told that Player B has chosen to cooperate, Player A increases his own rate of cooperation from 36 to 61 percent. Player A is willfully choosing to earn $5 instead of $10, when the supposedly rational thing to do would be to defect.

If you were going to play the game repeatedly with the same player, such a choice might be consistent with the axiom of self-interest. By using your choice to create a reputation as a cooperator, you can hope to earn $5 in future rounds of the game, rather than have your partner start defecting, thus leaving you with less over time. But in the studies I've described, the game is a one-time occurrence, and thus creating a reputation has no benefit. The only reasonable explanation is that in addition to being self-interested, we are also interested in the welfare of others as an end in itself. This, along with self-interest, is part of our basic wiring.

When you are playing the Prisoner's Dilemma and you have been told that Player B has already chosen to cooperate, your decision to cooperate implies that you care more about Player B's earning $5 rather than nothing than you care about earning $10 for yourself instead of $5. Given that you have not met (and will never meet) Player B, this is pretty remarkable. Would you have guessed that the typical stranger passing you on the sidewalk would engage in this kind of selfless behavior toward you? How about strangers in the most remote parts of the world? A large international collaboration examined fifteen preindustrial societies, from the foraging Au of Papua New Guinea to the farming Shona of Niger-Congo, and found that people made decisions counter to their own self-interest in each society. People around the world are willing to get a little less so that a stranger can get a little more.

Assuming for the moment that this behavior is not irrational, do people really prefer to see others do well? Or do people feel obliged to cooperate? After hearing the Golden Rule ("Do unto others as you would have them do unto you") for the umpteenth time, perhaps people feel that they are expected to treat others well, whether they want to or not. Perhaps such people believe that if they violate this rule, others will think less of them, and so they capitulate. This account is in keeping with scientist and philosopher Richard Dawkins's counsel that we "try to teach generosity and altruism, because we are born selfish."

Perhaps looking to the brain can help sort this out. We know what the brain looks like when we are complying with a social norm, and we know what the brain looks like when we are choosing based on real preferences. The former involves lateral parts of the prefrontal cortex (that is, the parts of the brain that let us inhibit our desires, among other things), whereas the latter involves the reward system in regions of the brain like the ventral striatum.

James Rilling, a neuroscientist and anthropologist at Emory University, conducted an fMRI study of subjects playing the Prisoner's Dilemma to find out what is going on in the minds of people as they cooperate or defect. Although staying true to a social norm might lead you to grudgingly cooperate more frequently, the reward system should reveal your real preference for the better financial outcome. Even if you cooperate 70 percent of the time out of a sense of obligation, a reward system that is selfishly motivated should respond more strongly on the remaining trials when you defect and earn more money for yourself.

In fact, the individuals in Rilling's study showed the opposite pattern. When the participants' partners chose to cooperate, more ventral striatum activity was observed in players when they too had chosen to cooperate, rather than defect. In other words, there was increased reward activity even though the players were earning less money for themselves. The ventral striatum seemed to be more sensitive to the total amount earned by both players, rather than to one's personal outcome. Moreover, the lateral prefrontal regions were not engaged in the study when subjects cooperated, suggesting that cooperating involves a real preference, not a sense of obligation.

The one hitch in Rilling's first study was that subjects played repeatedly with the same partners. Here, reputation building could have played a role such that early decisions to cooperate might have triggered the reward system as one considered how this strategy would yield larger rewards later on. But Rilling published another study a few years later in which only a single game was played with each partner, ruling out long-term strategies like reputation building. Nevertheless, he got the same results—mutual cooperation produced the greatest activity in the ventral striatum. Rilling also conducted trials in which the person in the scanner was informed that the game was being played with a computer opponent. In this case, mutual cooperation did not activate the reward system. Our reward system selectively responds to teaming up with other people, even if we earn less money in the process.

Our theory of "who we are" suggests that we cooperate in order to ultimately achieve a better end for ourselves. Once again we see this theory of human nature is misguided because it doesn't take into account the social motives that sit alongside our more familiar selfish motives. Mutual cooperation activates the reward system as an end in itself.

Mooting Altruism

In Isaac Asimov's book *The End of the Eternity,* a reality-altering time-traveler named Andrew Harlan falls in love with a woman from the future, Nöys. Knowing that her existence will be obliterated by the next change he is required to make, he hides her in a far distant future century where she will be unaffected. After he reveals his actions to her, and acknowledges that these actions constitute a great crime among his fellow time-travelers, she is shocked that he would risk his career for her. "For me, Andrew? For me?" she asks. To which he replies, "No, Nöys, for myself. I could not bear to lose you."

Are seemingly selfish acts that we observe ever really altruistic?

Historically, the question has been easy to pose and just as easy for skeptics to dismiss. *Altruism* is defined as helping others in such a way that the long-term material outcome of helping another is believed to have overall negative consequences for the helper. When Michael Ghiselin wrote, "Scratch an 'altruist' and watch a 'hypocrite' bleed," the implication is that "on closer inspection, . . . acts of apparent altruism are really selfishness in disguise." Perhaps the person who receives help will reciprocate directly. Or the person offering help will be seen in a more beneficial way in the eyes of others, allowing him or her to gain more later. We all wonder at times what people hope to gain from their seemingly altruistic behavior.

Understanding other people's psychological motivation is tricky because it's typically their word versus yours. Let's say John agrees to switch places with Elaine, who is receiving shocks as part of an experiment. Earlier John was not getting shocked, and now he is. Once John takes Elaine's place, Elaine leaves the experiment, never to be seen again. Surely John's act must be altruistic.

Psychologist Daniel Batson showed that there may be a hidden selfish motivation at work in John's willingness to switch, just like the protagonist in Asimov's story. Batson conducted ingenious studies in which one person (the observer) had to watch another person (the victim) receive painful shocks. The victim was clearly very bothered by the shocks and at one point asked if the shocks could be stopped. The experimenter then asked the observer if he would take the victim's place and receive the remainder of the shocks. Some observers were given the choice of either switching places or continuing to watch the victim receive shocks. Other observers were given the choice of either switching places or going home (without watching any more of the shocks). Those who would have to stay and continue to watch were much more likely to switch places with the victim than were those who could go home if they declined to switch places. In other words, if it is easy to escape the unpleasant situation, people do, but if it is hard, people decide that doing "the right thing" is better than having to watch the other person endure the shocks. Their willingness to let the victim continue to receive shocks, as long as they won't have to watch it happen, revealed that their motive was not purely altruistic.

But this study had a twist. Two other groups of observers were given the same choices—switch/stay or switch/leave. But these observers had been induced to feel empathy for the victim before the shock procedure began. The empathizing observers were very likely to switch places when the alternative was to stay and watch the victim receive more shocks. However, unlike the previous participants, the empathizers were also likely to switch places with the victim even when the alternative was leaving without watching any more shocks being given. In fact, the empathizers who had the option of escaping the situation were the most likely (91 percent) of any group in the study to agree to the switch. One has to conclude that the empathizers really were motivated by concern for the other person and not just whether they had to continue watching the other person receive shocks. These results imply that empathy is a catalyst for altruistic behavior, an idea I will return to in Chapter 7.

In considering whether altruistic behavior is really selfless, it is useful to consider the question of why we like to have sex. We can think about the motivation to have sex on at least two levels. First, there is an evolutionary motivation for us as a species to have sex because it leads to reproduction. Those individuals in our evolutionary past with a greater propensity to have sex—with a stronger sex drive—were more likely to reproduce and pass on their sex-preferring genes to their descendants. Yet the urge to reproduce is not the only or even the primary motivation why we as individuals have sex. No one is more sex obsessed than teenagers, and yet reproduction is usually the last thing on their minds. Indeed, fear of pregnancy is a strong deterrent against teen sex. Most people have sex because it feels good, physically and emotionally. The evolutionary motivation might be reproduction, but our psychological motivation is pleasure. Those who find sex more pleasurable are more likely to reproduce, often accidentally, and pass on those genes for enjoying sex.

This same analysis applies to altruistic behavior. Although a group of individuals may have a higher chance of passing on their genes if they cooperate and support one another, the psychological mechanism that motivates us to selflessly help others may be the intrinsic pleasure that we experience when we do it. If helping others gives us pleasure, what some call the *warm glow* of altruistic behavior, is this selfish or not? When we observe seemingly altruistic acts, we tend to look for the hidden selfish motive—some material benefit that the person will get in the long run to the ultimate disadvantage of others around him. In our search to uncover the selfish root of a behavior, we are unlikely to think, "He's just helping out because it makes him feel good. I bet he will continue to help others without expecting us to do anything for him in return. What a selfish bastard!" Yes, there is a sense in which we can characterize such behavior as selfish, but it is not the kind of selfishness that seems morally questionable.

As the Dalai Lama advises, "If you would like to be selfish, you should do it in a very intelligent way. The stupid way to be selfish is the way we always have worked, seeking happiness for ourselves alone and in the process becoming more and more miserable. The intelligent way to be selfish is to work for the welfare of others" because doing so is intrinsically pleasurable.

The Prisoner's Dilemma studies were the first to demonstrate that the brain's reward system responds to valuing the outcomes of others, in addition to one's own. One could argue that the study did not go far enough to prove the case, because when participants chose to cooperate, they were still getting paid, just not as much as they would have received if they had defected. But a more recent study provides even more compelling evidence that our reward system is sensitive to the welfare of others.

Jorge Moll and his colleagues at the National Institutes of Health ran an fMRI study looking at the activity in the brain when we're giving to charity. Individuals in the scanner were asked to make a series of decisions that involved financial outcomes for themselves and for a charitable organization (different decisions involved different charities). On some trials, individuals were asked whether they would agree to receive $5 for themselves with no consequences for any charity. Not surprisingly, individuals were very quick to accept this kind of reward. On other trials, individuals were asked if they were willing to give up some of their winnings (for example, lose $2) so that a charity would receive $5. Amazingly, as a group, the individuals in this study showed even greater activity throughout the reward regions of the brain when they made the choice to give away some of their own money to help others, compared to when they received money with no strings attached. Our supposedly selfish reward system seems to like giving more than receiving.

Eva Telzer, Andrew Fuligni, and I replicated this finding with what you might expect to be some of the most selfish people on the planet: teenagers. Instead of mentioning a charity, we asked teenagers to make costly donations to their own families. We told the teenagers, as well as their parents, that as a precondition of being in the study, any money given to the family must not be spent on the teenager who donated it. The majority of these teenagers reported taking pleasure in helping their families in daily life; they also showed increased reward system activity when donating their money to their families.

Along similar lines, Tristen Inagaki and Naomi Eisenberger examined supportive behavior between boyfriends and girlfriends. The women in the relationships were lying in the MRI scanner while their boyfriends sat next to them, just outside the scanner. On some trials of the experiment, the boyfriend would receive an electric shock and on others he would not. In both cases, the girlfriend in the scanner knew what was happening to him. On some trials she was instructed either to hold his arm with her hand or to hold a small ball. Physical contact with one's partner might be expected to be more rewarding than holding a ball, and, sure enough, this was the case. What was more surprising was that the reward system of the girlfriend showed the most activity when she was touching her partner during trials when the boyfriend was being shocked. On these trials, providing support through physical contact when the girlfriends knew their boyfriends were likely distressed was more rewarding than touching their boyfriends when no support was needed. Providing social support, even when doing so puts us in closer contact with someone else's distress, is reinforced in our brains. It feels good to help those we care about. Typically, when we think of the benefits of having good social support networks, we imagine ourselves being the beneficiary of support from others. But this finding suggests that our support of others could contribute significantly to our well-being.

We humans are complicated creatures. We are unquestionably self-interested. Adam Smith, one of the founders of modern economics, was astute when he wrote, "It is not from the benevolence of the butcher, the brewer, or the baker that we expect our dinner, but from their regard to their own self-interest." They help put food on our table because by charging us, they are able to put food on their own. Yet he was arguably even more wise when he suggested, "How selfish soever man may be supposed, there are evidently some principles in his nature, which interest him in the fortune of others, and render their happiness necessary to him, though he derives nothing from it, except the pleasure of seeing it."

We tend to think of rewards as material things (food, shelter, iPhones), and we think of those things as having objective value. Ten dollars is always better than five, and five is always better than zero. But material rewards are rewarding only because our brains evolved to experience those things as rewarding. We are also built to take pleasure in cooperating and helping others. We can call it "selfishness," but if we do, the notion of selfishness ceases to be a bad thing. The neuroscience of cooperation and charity eliminates the typical question of altruism ("Are we ever altruistic?") and replaces it with two new questions: Why are we evolved to enjoy being altruistic? and Why don't we realize that being altruistic can be intrinsically rewarding? Let's take these questions in order.

Why Are Social Rewards Rewarding?

As we've seen, there are two kinds of social rewards—the social rewards we receive when others let us know they like, respect, or care for us and the social rewards we receive when we care for or treat others well. It is no accident that this parallels the two sides of the mother-infant relationship. Having strangers tell us they like us is pleasurable, in part, because we humans have generalized the positive feelings of being cared for by our mothers. Many mammalian species have shown opioid-linked pleasure responses in the brain while being groomed by their mother or peers. But in humans most of our grooming is verbal rather than physical. When others spend time verbally grooming us, it is a sign that we are safe and cared for. And given our long period of immaturity, this is an incredibly reinforcing signal to receive.

That being treated well by others is rewarding isn't surprising. We know it feels good to be liked and cared for. It's a sign that others will include us when there are material goods to be divvied up between members of the group. But how do we explain the fact that we are sometimes motivated to help others, even complete strangers, when there is no material benefit for ourselves? How do we explain truly altruistic sentiments? The best answer may have something to do with an evolutionary change in parental caregiving.

Mammalian mothers of all stripes are jump-started into caregiving mode as a function of the birth of their offspring. Rats begin to bond and groom their offspring within a few days of their pups' being born; mother sheep bond to their offspring within two hours of birth; and humans begin mentally bonding months before the baby is even born. In all cases, the neuropeptide *oxytocin* is a critical driver of our caregiving motivations. Oxytocin's primary physiological contribution is to facilitate labor during the birthing process and to promote the flow of milk during breastfeeding. Within the brain's reward system, oxytocin also motivates us to approach our infants to support their well-being, and it diminishes the personal distress we ordinarily feel at approaching someone else in distress.

Thus, the two kinds of social rewards depend on different kinds of neurochemical processes. Being cared for promotes opioid-based pleasure processes in the brain. In contrast, the effects of oxytocin may be better characterized as modifying the dopaminergic processes that promote approach behavior. We reach for the Snickers bar because dopaminergic signals tell our brains that if we eat the Snickers bar, we will enjoy it. In simple terms, we gravitate toward things the brain has learned to associate with dopaminergic release. Mammalian brains are loath, however, to approach strangers because they may represent a threat. And to a rat, a newborn pup really is a stranger. Mammals are thus in a bit of a bind because, on the one hand, their offspring are strangers that we are built to avoid, and on the other hand, caring for our young is essential for their survival. Oxytocin appears to alter the dopaminergic response of mammals to their own infants, tipping the balance from avoidance to approach.

It has been suggested that oxytocin is a love drug or a trust hormone, but I prefer to think of oxytocin as the *nurse neuropeptide*. After college, I spent a year working as a clerk on a surgical unit at St. Peter's Hospital in New Brunswick, New Jersey. I worked with nurses every day and the work they do is extraordinary. Their work is very hard and not so obviously rewarding—much like parenting can be. Each day, they deal with patients and family members who are at their worst. And unlike the rest of us, whose stomachs turn at the sight of bodily fluids, and whose eyes roll up into our heads at the wounds that must be dressed, nurses rush in and do what needs to be done. They don't do it because they love the patients or trust them. Often they barely know the patients. They do it because they are motivated to help, as an end in itself. Oxytocin turns the rest of us from zeros to heroes when it comes to caring for our own children. Nurses do it for everyone every day.

In animals, prosocial sentiments toward one's offspring have been associated with higher levels of oxytocin modulating reward responses in the ventral striatum and ventral tegmental areas of the brain—both part of the reward system. One account suggests that oxytocin released in the ventral tegmental area leads to the release of dopamine in the ventral striatum region associated with increasing our motivation to seek out a reward. Fearlessness appears to be influenced by oxytocin interactions within the septal region, adjacent to the ventral striatum. Both oxytocin and the septal region of the brain are involved in diminishing the physiological indicators of distress, which may facilitate helping someone else even when the situation is distressing or gross. In other words, when we see someone in need, say, someone with a bloody wound, oxytocin may simultaneously increase the reward value of approaching that person and decrease the distress we might have over being near someone else in distress.

Although there are great similarities in how oxytocin promotes care for offspring across mammalian species, oxytocin has different effects on how primates and nonprimates treat strangers. In nonprimates, increased oxytocin is associated with increased aggression toward strangers. This is generally understood in terms of mothers' protecting their infants from unknown threats. A mother sheep will attack an unrelated baby lamb that tries to nurse from her. But when the oxytocin processes are blocked, the mother sheep will allow the unrelated lamb to nurse. Thus, in nonprimates, oxytocin promotes direct care of one's own offspring, including protecting them against others. This ensures that the mother's limited resources are spent only on those offspring that will pass on her genes to future generations.

Both the caring- and aggression-related effects of oxytocin have been demonstrated in humans as well. Administering oxytocin has been shown to increase generosity when people play behavioral economics games like the Prisoner's Dilemma. On the flip side, psychologist Carsten De Dreu in the Netherlands has demonstrated in multiple studies that administering oxytocin leads to more aggressive responses to members of other ethnic groups in the Prisoner's Dilemma.

While oxytocin can promote ingroup favoritism (that is, toward groups that one is a part of) and hostility toward those who are not part of one's ingroup, the dividing line between friend or foe differs in a crucial way between primates and other mammals. In nonprimates, oxytocin leads individuals to see all outsiders as possible threats, thus enhancing aggression toward them. In contrast, humans divide others into at least three categories: members of liked groups, members of disliked groups, and strangers whose group affiliations are unknown. Administering oxytocin in humans facilitates caregiving toward both liked group members *and strangers,* but it promotes hostility toward members of disliked groups.

Oxytocin in humans helps to promote altruistic tendencies not toward one's own group—because that isn't altruism in the strongest sense of the word—and not toward members of disliked groups. But oxytocin can increase our generosity toward complete strangers, which is quite magical, as strangers who start with a positive bias toward one another can do great things together, such as building houses, schools, and other institutions that support a society.

Why Don't We Know?

If you scanned my brain while I ate a scoop of salted caramel ice cream, you would undoubtedly find increased activity throughout my brain's reward system. Or you could save a lot of money on fMRI scans and just ask me if I love salted caramel ice cream. When it comes to ice cream, our conscious experiences and our brains tell the same story. So why isn't it the same with social rewards? Why doesn't it seem like being treated fairly would feel good? Why don't we recognize that there is something intrinsically rewarding about helping others that does not depend on believing one will benefit materially? Research suggests the reason is that we feel compelled to tell everyone how selfish we are—even if we aren't selfish.

I was recently at a meeting of the social psychologists in my department—professors and graduate students. The area chair made a point of thanking Kelly Gildersleeve, a graduate student who had spent a lot of time over the summer streamlining a bureaucratic process by moving it to the Internet. When everyone in the room heartily applauded her effort, Kelly blushed and blurted out that she was going to benefit from the streamlining as well, so she would be getting something out of it. It was a complete lie. The time she put in will never be offset by the savings she will get in her last year of graduate school. Kelly told me later that even as she heard herself saying this, she knew it wasn't true. But she said those selfish-sounding words anyway. Kelly helped because she is a kind and thoughtful person who saw a problem and knew she could help. Kelly helped because it is intrinsically rewarding to help those around you. For some reason, though, when someone asks us why we help, we often find ourselves saying things that makes us sound more selfish than we are.

Dale Miller, a social psychologist at Stanford University, has identified the root cause of this faux-selfish behavior. The theorizing of Hobbes, Hume, and other intellectuals who claim that self-interest is the source of all human motivation has produced a self-fulfilling prophecy. Their theory and everyone who repeats it have affected how the rest of society behaves. Because we have been taught that people are self-interested, we conform to this cultural norm to avoid standing out.

Miller has shown in multiple experiments that we assume others are far more self-interested than they really are. In one study, he asked individuals what percentage of undergraduates they thought would agree to give blood for $15 and what percentage would agree to give blood if there were no financial incentive. Respondents estimated half as many people would give blood for free as would for the money (32 versus 62 percent). But in measuring actual volunteer rates, he found that those who were offered no money agreed to give blood 62 percent of the time, only slightly less often than those who were paid (73 percent).

Because of these mistaken assumptions about everyone else's selfishness, we tend to avoid appearing altruistic ourselves. We don't want to appear to be boasting or come off as a goody-two-shoes. If you believe that people in general don't think altruism exists, then claiming your own actions are altruistically motivated feels like putting yourself on a pedestal. For this reason, when people are asked why they have engaged in prosocial behaviors, they tend to ascribe their actions to self-interest ("I volunteer because I'm bored, and it gives me something to do"). When we regularly hear other people giving selfish-sounding reasons for their altruistic behavior, it only serves to bolster our belief that all behavior is self-interested, which in turn makes us less likely to admit our own altruistic motives. The cycle is self-reinforcing, becoming more and more ingrained over time.

The irony of this was brought home in another of Miller's studies. People were approached to donate to a charity. People who were asked to simply donate found it hard to generate a self-interested explanation for helping the charity. Other people were informed that they would receive a small candle in return for their donation. The candle created an *exchange fiction*, allowing people to say, "I didn't donate to help. I was buying a candle." As expected, people were more likely to donate when they would get a candle in return compared to when no candle was offered. They also donated much more money under these conditions. Ironically, getting a trinket in return allows us to cover our generosity with a nonaltruistic account and thus frees us to act more altruistically.

Alexis de Tocqueville, the French scholar who wrote the first great book about the United States, *Democracy in America*, in 1835, noted his surprise when it came to Americans' thoughts about their own good deeds: "The Americans . . . are fond of explaining almost all the actions of their lives by the principle of self-interest. . . . In this respect I think they frequently fail to do themselves justice; in the United States as well as elsewhere people are sometimes seen to give way to those disinterested [that is, not self-interested] and spontaneous impulses that are natural to man; but the Americans seldom admit that they yield to emotions of this kind."

The fact is, we are full of both selfish and unselfish motives. And this is no accident. Mammalian brains are wired to care for others, and among primates this caring extends to at least some non-kin, even when there is no material return on the investment. Because of the way our brains are wired, eating a delicious piece of cake is enjoyable whether we are hungry or not. Similarly, helping others feels good whether we expect something in return or not.

Just imagine what things would look like if we were taught about this in school and we understood that altruistic helping is just as natural as being selfish. The strange stigma associated with altruistic behavior would be lifted, perhaps engendering far more prosocial behavior.

A Lifetime of Pain and Pleasure

In this chapter and the previous one we have looked at two of the major evolutionary motivational tools that work together to ensure that all mammals are concerned with their social world. Pain and pleasure are the driving forces of our motivational lives. The animal kingdom is full of species that successfully avoid threats that may cause harm, and they are drawn to potential rewards that can help them survive and reproduce. It isn't surprising that mammals are built to avoid predators or to remember where they found the cheese in the maze last time.

What is surprising is that these basic pain and pleasure motives have been co-opted to serve our social lives as well. The single most important need of an infant mammal is to be continuously cared for by an adult. Without this, all other needs of the infant go unmet, and it will die. Creating ways to keep us connected is therefore the *central problem of mammalian evolution*. By making threats to our social connection truly painful, our brains produce adaptive responses to these threats (for example, an infant's crying, which gets a caregiver's attention). And by making the care of our children intrinsically rewarding and reinforcing, our brains ensure that we will be there for our children even before we are needed.

Oftentimes there are unintended consequences of evolutionary adaptations. Did the need to be socially connected and the pleasure we take in caring for others contribute to the evolution of romantic relationships that extend beyond simple procreation? These pain/pleasure responses may have evolved for the purpose of infant caregiving, yet they stay with us for a lifetime, radically shaping our thoughts, feelings, and behaviors till the end of our days. The downside to these social motivations is that they can have truly harmful consequences when they go unsatisfied. The severing of a social bond—whether it's the end of a long-term romantic relationship or the death of a loved one—is one of the greatest risk factors for depression and anxiety. Although adults can survive with unmet social needs far longer than with unmet physical needs, our social bonds are linked to how long we live. Having a poor social network is literally as bad for your health as smoking two packs of a cigarettes a day.

The social motivation for connection is present in all of us from infancy. It is a pressing need, with a capital *N*. The evolutionary fallout from the presence of these social needs is a major advantage to those who are able to minimize their social pains and maximize their social pleasures. Building and maintaining social networks is no easy feat. Just watch any reality show, from *Survivor* to MTV's

Real World. Fortunately, evolution has given us not one but two brain networks that help us to understand those around us and to work more cohesively with them. Connection is the foundation on which our social lives are founded, but evolution was far from finished, making sure we would make the most of our social lives.

Part Three

Mindreading

CHAPTER 5

Mental Magic Tricks

Most of us believe that a coin flip can resolve nearly any stalemate fast and fairly. The ancient Romans called this *navia aut caput,* referring to the ship and the head on the two sides of their coins. Flipping a coin seems like a reasonable way to resolve such standoffs because it appears just as likely to land on heads or tails. Except that it isn't. A few years ago, a group of medical residents were each asked to flip a coin 300 times under rigorous testing conditions and to try to make the coin come up heads on each flip. These were not gamblers or con artists, and they were not given much time to practice. Nevertheless, each resident was able to flip more heads than tails. One resident turned more than 200 heads, for a hit rate of 68 percent—far above random chance. Statisticians from Stanford University analyzed the physics of coin tossing and determined that without any mischievous intent on the part of the flipper, a fair coin will tend to land facing the same way it started. The same-side advantage is small (51 to 49 percent); but if true, who would ever agree to a coin toss again?

When San Francisco 49er Joe Nedney heard about the coin flipping results, he suggested a switch to rock-paper-scissors to decide which NFL team would kick off or receive the ball each game. How many children have been doomed to retrieve the ball from the scary neighbor's yard by a lost round of rock-paper-scissors? If coin tossing is out and rock-paper-scissors is the fair way to break a deadlock, then Bob Cooper must be the luckiest guy on the planet.

In 2006, he defeated 496 contestants to claim the title of Rock-Paper-Scissors World Champion.

As most people know, the game is simple. Two players simultaneously reveal a hand gesture indicating rock, paper, or scissors. Rock crushes scissors. Scissors cut paper. Paper covers rock. Players have three options, and each option can beat one other and is beat by one other (if both players select the same gesture, it's a tie, and they go again). In the final match of the Rock-Paper-Scissors World Championship, Cooper, a sales manager from London, went up 5 to 2 by throwing rock to his opponent's scissors. He needed one more for the win. In the next round they both threw paper. Then rock. Then scissors. Three tie rounds in a row. Finally, in the fifteenth round of their match, Cooper threw scissors over paper and was crowned the king of rock-paper-scissors.

The match can be viewed on YouTube, and the first comment appearing below the video goes for the high sarcasm of "I can't wait to see the coin flip championships." To the uninitiated, rock-paper-scissors seems random, with each party having an equal chance of winning. If you believe that, I know some rock-paper-scissors experts who would love to schedule a high-stakes match with you. The best of the best are mindreaders; they know what you will play before you do. Novices have clear tendencies that can be exploited by opponents. For instance, men who are new to rock-paper-scissors tend to start matches by throwing rock more often than paper or scissors, possibly because rocks are associated with strength. Another tendency is to change gestures after the same one has been thrown two rounds in a row. Players with more experience can counter these novice moves, dramatically increasing their likelihood of winning.

Of course, in competition play there are no novices. Experienced players carry out a series of complicated attacks and counterattacks. After being crowned world champion, Bob Cooper told a reporter that the essence of rock-paper-scissors is about "predicting what your opponent predicts you'll throw." It's about getting inside your opponent's head and manipulating what he believes you will throw and understanding how he will use that information to counter you, so that you can in turn throw a gesture that will counter him. It's all about mindreading.

Marcus du Sautoy, a professor of mathematics at Oxford University, tried the one strategy that does not depend on mindreading and would seem invincible to the mindreading efforts of others. He decided each throw entirely randomly, using successive digits in the number pi (3.1459 . . .) to determine his next throw. By not strategizing at all from round to round, he took away his opponents' ability to manipulate him. But although he had some luck with his "can't lose, can't win" strategy, he was no match for Bob Cooper. Cooper beat him eight straight times. Rather than relying on statistical knowledge, Cooper was probably detecting subtle facial expressions and body language that gave away du Sautoy's next move. More mindreading.

Everyday Mindreading

Franz Brentano is a little-known German philosopher who is the forefather of some of the most important philosophers and psychologists of the twentieth century. He trained Edmund Husserl, who later trained Martin Heidegger, one of the giants of modern phenomenological and existential philosophy. He also trained Carl Stumpf, who trained the first Gestalt psychologists ("the whole is greater than the sum of its parts"), and Kurt Lewin, who is considered one of the founders of social psychology in the United States (having left Germany at the start of World War II).

In 1874, Brentano published a long forgotten text, *Psychology from an Empirical Standpoint,* that along with Wilhelm Wundt's enormously influential *Principles of Physiological Psychology,* published in the same year, were the first modern texts on psychology. Brentano argued that the central fact of human psychology is that our thoughts are "Intentional." Brentano's meaning is derived from Aristotle and twelfth-century scholastic philosophers who had discussed the "intentional inexistence" of objects. Essentially, *Intentionality* refers to the fact that we have thoughts, beliefs, goals, desires, and intentions about other things. Our thoughts can be about objects in the world or about imaginary entities like wizards at Hogwarts, or even about other thoughts, but thoughts always extend beyond themselves to refer to something else. Nothing else in the known universe has this intrinsic characteristic of "aboutness" (for example, rocks aren't "about" anything; they just are).

It took another half-century after Brentano before the corresponding central fact about our social minds was identified: we possess the capacity or, more accurately, the inescapable inclination to see and understand others in terms of their Intentional mental processes. When we see others, we want to know what they are thinking about and how they are thinking about it.

To first demonstrate this penchant for everyday mindreading, Fritz Heider showed people a short animation of two triangles and a circle moving around, and then he asked people what they saw (see Figure 5.1). Here's what people didn't see: two triangles and a circle moving around. Instead, people saw drama. *"The big triangle is a bully that is picking on the small triangle and circle, who are running scared but then figure out how to trick the big triangle and escape."* Or *"The big triangle is a jealous boyfriend of the female circle, and he is angry because he caught the circle flirting with the small triangle."* Everyone saw thoughts, feelings, and intentions in these shapes that clearly had none—shapes don't have minds! We see thinking, feeling minds everywhere around us: we treat our computers, cars, and even the weather as if they have minds of their own. This overgeneralized tendency to see minds behind events in the physical world presumably evolved to make sure we do not accidentally overlook the actual minds of other people. Actual minds are hidden, after all. It would be awfully easy to miss them if we weren't built to notice them.

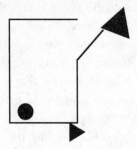

Figure 5.1 Heider and Simmel's Fighting Triangles.
Adapted from Heider, F., & Simmel, M. (1944). An experimental study of apparent
behavior. *American Journal of Psychology*, 57, 243–259.

In 1971, a century after Brentano's pronouncement, the philosopher Daniel Dennett codified our tendency to see others in terms of minds guiding behavior. Dennett suggested that regardless of whether we were justified in assuming that other minds exist, we are built to assume that others are Intentional creatures. Dennett referred to this as taking the *Intentional stance.* It is because our own minds think about the minds of others (who in turn think about us) that we can have the high comedy of Daffy Duck's "I know that you know that I know" standoffs with Bugs Bunny. While Bugs and Daffy might be overdoing it, this kind of interaction is one of the primary reasons why societies are able to work cooperatively to build soccer leagues, schools, and skyscrapers.

The Intentional stance is so ubiquitous and so easily adopted in daily life that it is almost impossible to appreciate what an achievement it is. We are all mindreaders. As you read this, you comprehend not just the marks on the page but thoughts that I had when I was writing them. Similarly, in writing these words, I have to be able to predict how these marks on the page will be experienced in your mind in order to make my thoughts more easily understood. How on earth would you go about building a machine that assumes others have these mental processes and can usefully take them into account? As impossible as this might sound, we do this all the time without even realizing it. Perhaps that was why it took so long for anyone to recognize we have this capacity. Like fish who have no idea that they are in water because they are surrounded by it, mindreading is so basic to who we are that we rarely notice it.

Just try to imagine how you would get through your day if you couldn't make sense of the minds of others or count on others to make sense of your mind. Consider the most trivial example. Whenever I take a flight home to Los Angeles, a shuttle picks me up to bring me to where my car is parked. As the shuttle approaches my terminal, I wave my hand, and the driver knows that I would like him to stop so I can get on. When he stops and opens the door, I know the driver's intention as well—he is inviting me to climb aboard. It is a simple transaction between two complete strangers. Yet without each of us having an accurate understanding of the psychological meaning of other people's behavior, we would be unable to pull off this innocuous interaction. Now consider a consulting team working with a company to develop new hiring plans or a math teacher instructing two dozen teenagers about the finer points of sines and cosines. In these cases, we must have exquisite insight into how our actions will be understood by those we are working with. The modern world would stop in its tracks if we no longer had this ability to understand or predict the minds of others. Our ability to think allows us to imagine great achievements, but without the ability to think socially and share our vision with others in a way that engages them, we would be left to our own devices to convert our vision to reality.

Psychologists have referred to this ability to understand that other people have thoughts that drive their behavior as having a *Theory of Mind,* and when people apply this ability, it is called *mentalizing* (that is, we mentalize when we think about the mental states of others). Just as scientists have theories that allow them to make predictions and draw conclusions based on evidence, we as human adults all operate as if we have a theory that others around us have a mind that responds in an orderly fashion based on a set of rules (for example, losing a game makes people sad, not happy). It is this signature achievement that allows us to coordinate our otherwise isolated thoughts with the thoughts of others to promote shared goals and cooperation.

Punch and Judy

Over the past three decades, Theory of Mind researchers have focused on two related questions: Who has it, and When do they get it? The *who* question is usually aimed at determining which nonhuman species, if any, share this Theory of Mind capacity with us. Are we humans alone on the planet in our ability to appreciate the minds of others, or like so many capacities, such as using tools, are the differences between us and the rest of the animal kingdom more a matter of degree? David Premack and Guy Woodruff were the first to take up the challenge of sorting this out. Chimpanzees are the closest living relative to humans, genetically speaking, so if any other animal were to have a Theory of Mind, they would be the most likely candidates. Premack and Woodruff worked with a chimpanzee named Sarah who could perform a pretty impressive trick. The researchers would show Sarah a video of a man engaged in some activity like trying to get a banana that was too high to reach. The video would be paused before the man had solved the problem, and Sarah would be given four photographs showing possible next steps for the banana-hunting man. Sarah could reliably pick the photo that indicated the right solution (getting a box and standing on it). Although that would be easy for you and me, it is remarkable that a chimpanzee could do it too. Premack and Woodruff suggested that the only way Sarah, a chimpanzee, could do this was to understand that the man was the kind of entity that could have desires and goals and that in this case he had a particular desire-goal combination: to satisfy his hunger by getting the banana.

So chimpanzees have a Theory of Mind? In the end, Sarah's feat was more of a debate starter than a debate ender. Dennett and others weighed in that impressive as this trick might have been, it may have reflected no more understanding than a parrot's being trained to ask a question based on conditioning, or solving a problem for oneself ("What would I do?") without thinking about the other's mind. Dennett (1978) proposed a more definitive *false belief task* based on the eighteenth-century comedic stylings of Punch and Judy:

> Very young children watching a Punch and Judy show squeal in anticipatory delight as Punch prepares to throw the box over the cliff. Why? Because *they know Punch thinks Judy is still in the box.* They know better; they saw Judy escape while Punch's back was turned. We take the children's excitement as overwhelmingly good evidence that they understand the situation—they understand that Punch is acting on a mistaken belief.

Dennett's critique led to the second dominant Theory of Mind question, which has focused on *when* humans demonstrate this ability during the course of their development. As Dennett's example implies, humans do appreciate false beliefs in others, but they aren't born with this ability. In the mid-1980s, a number of researchers converted Dennett's Punch and Judy thought experiment into a real one. The best-known Punch and Judy variant is known as the *Sally-Anne task.* Two puppets, Sally and Anne, are both seen, along with a basket and a box. Sally puts a marble into the basket and then leaves the stage. While Sally is away, Anne moves Sally's marble from the basket to the box. When Sally returns, the child watching the performance is asked where Sally will look for her marble. The trick here is that the child watching this mini-drama has a true belief about where the marble is, whereas Sally has a false belief. Sally still thinks the marble is in the basket where she left it, but she's wrong. If children have an egocentric view, believing that everyone knows what they know, they will say Sally will look in the box. However, if they can appreciate that others can have beliefs that differ from their own and can have beliefs that do not line up with reality, they will be more likely to say that Sally will look in the basket. The results from many studies provide strong converging evidence. Three-year-olds are lousy at the test, and five-year-olds are great at it.

As new and different tests are devised, younger and younger children also show some evidence of this sort of social skill. Chimpanzees show evidence of precursors of this ability, but no evidence unambiguously demonstrates that they can cross the threshold of thinking about the false beliefs of others. Humans may be alone in the universe when it comes to their ability to thoroughly appreciate the nature of others' minds.

Why is it so amazing that children think about the mental states of others? Because mental states are invisible. Have you ever seen a thought, feeling, or desire? Yet somehow, we learn to infer that these invisible entities in other people's heads are leading them to do the things they do. When we see a rock rolling down a hill, we don't think, "It wants to get to the bottom." But when we see a person running down a hill, we do.

Over time we develop a very complex theory of how different situations and outcomes are likely to affect a typical person's thinking and how that person will subsequently behave. If Bill and Ted are best friends, but Ted starts spending a lot more time with George, we know how Bill will feel (neglected, jealous) and how he will respond (either trying to integrate George into the group, making a stable triangle of friends, or competing with George for Ted's affection). I suspect I could describe just about any situation and you would feel confident about how a typical person would react. It is this ability to consider the mental reactions of those around us, to imagine these reactions in advance, that allows us to increase our exposure to social rewards and minimize the experience of social pain. If you can predict that the e-mail you are about to send to someone will lead that person to reject you, you can edit the e-mail to get your point across more tactfully. We do this countless times, in large and small ways, each day. We use our capacity for mind-reading to support our motivation for connection.

A System for General Intelligence

How is it that we perform this trick of understanding other people's minds? One of the earliest accounts focused on our general ability for abstract reasoning and effortful thinking supported by the prefrontal cortex. Logical reasoning comes in two flavors: deductive and inductive.

In deductive reasoning, we assess what *must* be the case if a set of premises is assumed to be true. Consider the following premises:

1. *If it rains, then the picnic will be canceled.*
2. *It is raining.*

If those two premises are true, then we must logically conclude the picnic has been canceled. This is an example of deductive reasoning, and this kind of *if-then* reasoning is central to our prodigious problem-solving abilities.

In contrast, inductive reasoning uses what has been true in the past to predict what will likely be true in the future. For instance, our belief that the sun will come up tomorrow is predicated on the assumption that the sun's having come up every day of our lives so far is strong evidence that it will continue to do so. Unlike deductive conclusions, sensible inductive inferences are not guaranteed to be true. There is nothing about past sunrises that actually guarantees more in the future, any more than the production of twelve *Friday the 13th* movies guarantees one more. Inductive reasoning produces conclusions that usually turn out to be true if the conditions of the world stay the same, which is why, so far, it has made sense to predict more sunrises and more *Friday the 13th* movies.

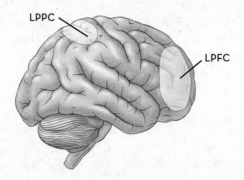

Figure 5.2 Lateral Prefrontal and Parietal Regions Associated with Intelligence, Reasoning, and Working Memory (LPFC = lateral prefrontal cortex; LPPC = lateral posterior parietal cortex)

Numerous neuroimaging studies have identified regions in the lateral prefrontal cortex and the lateral parietal cortex, also called the *lateral frontoparietal cortex*, that are more active when we are engaged in either deductive or inductive reasoning, compared with a task that involves neither kind of reasoning (see Figure 5.2). Some studies of the brain have shown differences between the two kinds of reasoning, but the neuroanatomical similarities are far more conspicuous than their modest differences.

More generally, these lateral frontoparietal regions support countless kinds of effortful thinking through a process known as *working memory*. Working memory is the psychological process commonly associated with mentally holding and updating multiple pieces of information. If I showed you a seven-digit number on a computer screen (8675309) and then asked you to remember that number for 10 seconds once it disappeared from the screen, working memory is what keeps it active in your mind. Similarly, working memory can be used to consider the relationship between things held in mind, like which of two numbers is larger.

To get a sense of how important working memory is to everyday functioning, consider your ability to read. By the time you got to the end of the previous sentence, you were using working memory to hold the beginning of the sentence in mind so that you could understand the complete train of thought. Imagine that you could process each word you read only at the moment you read it and then it was gone from your mind. You would never have the context of earlier parts of a sentence available to clarify the meaning of the later parts. Countless fMRI studies of working memory have implicated the lateral frontoparietal regions of the brain. When the degree of working memory load increases (for example, rehearsing a five-digit versus a seven-digit versus a nine-digit number), so does activity in these brain regions. It makes sense that some of the same regions would be involved in logical reasoning and in working memory: engaging in logical reasoning involves holding pieces of information in mind and comparing them, and this is the kind of thinking that working memory helps us with.

Working memory and reasoning abilities both overlap with our concept of general intelligence. People who can hold more information in mind and reason about that information effectively are seen as more intelligent than others who cannot. It probably won't come as a surprise, then, that studies of the neural bases of intelligence typically point to the same lateral frontoparietal regions involved in working memory and reasoning. People who score high on tests of fluid intelligence, a test of active thinking ability, have these regions turned on more when performing tasks that involve effortful or active thinking.

Given that all these kinds of thinking and reasoning recruit the same regions in study after study, a natural first hypothesis about Theory of Mind would focus on these regions of the brain. The lateral prefrontal cortex is the brain's all-purpose abstract reasoning device that helps you do your taxes, play chess, and remember the phone number you saw on an infomercial long enough to order your Flowbee. If it supports reasoning in general, why shouldn't it support reasoning about other minds? Just like general reasoning, the structure of social reasoning can be deductive or inductive.

Let's look at the Sally-Anne false belief task:

1. *Sally put the marble in the basket.*
2. *Sally does not see Anne move the marble to the box.*

Those premises yield the logical conclusion that Sally does not know the marble has been moved and therefore she holds an incorrect belief about the marble's location. This is standard deductive reasoning, and when we solve this problem, it feels no different to us than other kinds of deductive reasoning. Similarly, we draw on our past experiences in social settings to make inductive predictions. For example, we have seen people feeling disappointed when they get a low grade on an exam. From those observations, we can predict how someone will feel in the future if they get a low grade. But as parsimonious as this explanation sounds, the idea that social thinking is just like nonsocial thinking turns out to be wrong—completely wrong.

A System for Social Intelligence

Surprisingly, even though social and nonsocial thinking are structurally and experientially similar, the brain typically handles these two kinds of thinking using very different neural systems. Chris and Uta Frith published an early neuroimaging paper showing this. Individuals in their study read three kinds of sentences. Some sentences went together to form a paragraph that required mentalizing to be understood. One of these paragraphs told a story about a burglar who dropped his glove while running past a police officer. The police officer yelled, "Hey, you! Stop!" so that he could give the glove back to the burglar, but the burglar wrongly assumed he had been caught and gave himself up. In order to understand the burglar's behavior, the reader needed to understand the burglar's false belief that the officer was yelling at him because the officer knew he had committed a crime. Other sentences in the study did not tell a story, were unrelated to one another, and were unlikely to invoke mentalizing (for example, "The name of the airport had changed," and "Louise uncorked a little bottle of oil").

Just as in other studies of basic reading comprehension, when subjects in an MRI scanner read the unrelated sentences, they mostly produced activity in lateral prefrontal regions associated with language and working memory. However, when the sentences were put together in a way that induced mentalizing, the lateral prefrontal regions were relatively quiet. Instead, a different set of regions, including the dorsomedial prefrontal cortex (DMPFC), the temporoparietal junction (TPJ), the posterior cingulate, and the temporal poles, were more active (see Figure 5.3).

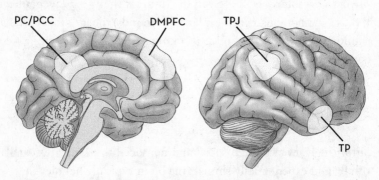

Figure 5.3 The Mentalizing System (DMPFC = dorsomedial prefrontal cortex; TPJ = temporoparietal junction; PC/PCC = precuneus/posterior cingulate cortex; TP = temporal poles)

Remember Heider's animated drama involving two triangles and a circle—the inanimate shapes that spontaneously elicit thoughts about the shapes' thoughts, feelings, and intentions? In another study, the Friths found that when people watched these animations, they produced selective activity in the DMPFC and the TPJ—just as in the previous study. However, individuals with autism, who have mentalizing deficits, showed weaker activity in these regions than nonautistic participants. So viewing geometric shapes that could be interpreted socially, without any specific instructions to do so, produces activity in regions involved in mentalizing. But these regions do not increase their activity in those who have trouble with mindreading in daily life.

One of my favorite mentalizing studies was run recently by psychologist Roberto Cabeza. His team took a more naturalistic approach to mentalizing, asking people to walk around wearing at chest level a camera that would automatically take pictures at regular intervals. At the end of this process, each person had hundreds of images of their mundane everyday experiences. Participants then went into the MRI scanner and saw the pictures in order. They also watched another individual's picture show. When looking at their own pictures, their experiences came back to them. But for the other person's pictures, they had to mentalize to imagine the experiences that would connect the dots between those pictures ("Where is this person going?" and "What is she trying to do?"). The regions associated with mentalizing (the DMPFC and the TPJ) were more active when viewing someone else's pictures compared with viewing one's own.

Across dozens of such studies conducted in the past fifteen years, two things have remained pretty constant. First, the DMPFC and the TPJ are almost always more active when people mentalize (with activity in the posterior cingulate and the temporal poles also showing up pretty regularly). Consequently, I refer to these regions as the *mentalizing system*. Second, heightened activity in the regions of the brain involved in working memory, nonsocial reasoning, and fluid intelligence are almost never observed in these studies. In other words, the neuroimaging findings are telling us something we could probably never have learned by just thinking about the inner workings of our minds: although social and nonsocial thinking feel like the same kind of process, evolution created two distinct systems to handle them.

Mentalizing by Default

This is not the first time we've encountered the mentalizing system. In its first appearance, back in Chapter 2, I referred to it as the *default network*. These regions that are involved in understanding the minds of others are largely the same regions that turn on whenever a person is given a moment of peace in the scanner, between cognitive tasks. These are the same regions that "turn on" when we dream. These regions start working together as a network from the day we are born. Earlier, I characterized these regions as helping to promote our intense interest in the social world. Having now seen the function of this network in terms of mentalizing, we have a much clearer picture of what this specialized network does for us.

Robert Spunt, Meghan Meyer, and I recently ran a study in order to piece together what is happening at rest and how it relates to our social focus on other people's minds. Previous studies have demonstrated that the default network and mentalizing network overlap anatomically; anyone looking at the two networks can see clearly that they are pretty much the same thing. The big question is whether the activity we see that is present during rest is really doing something social and whether that something serves an important purpose. Perhaps the network is doing something different at rest than it does during a mentalizing task. This has been unclear. To date, most accounts trying to look at the function of the default network have suggested it is mostly something that gets in the way, making us more error prone.

I suggested that the default network might provide the brain with thousands of hours of practice processing social information. If this is the case, people who have been turning the default network on more strongly all those years should now be better at social thinking—after all, more practice should lead to better results. As a small step toward examining this, Bob, Meghan, and I measured how strongly a group of individuals activated the default network during rest. If someone strongly activates this network now, the activation might reflect a history of having strongly activated this network at rest in the past, leading to enhanced mentalizing skills in the present. To test this idea, we correlated the strength of each person's default network activity with their mentalizing performance in a separate task.

Those who activated the DMPFC more while resting in the scanner were significantly faster when performing the mentalizing task later on. In fact, folks who activated the DMPFC the most were 10 percent faster than those who activated this region the least. Imagine being 10 percent better in every social interaction you have. It's like being a chess move ahead all the time. This is the first link between default network activity and actual social thinking. But without having studied these people over time, we couldn't be sure that the default network activity during rest was what was causing the enhanced social thinking. So we performed a second, more targeted, set of analyses.

Our second hypothesis was that the default network affects our moment-to-moment readiness to think socially. I discussed back in Chapter 2 the idea that default activity during rest might serve as a prime, getting us ready to see whatever comes next in terms of its social, rather than physical, aspects. More specifically, the default network may prepare us to see the actions of others through the lens of mentalizing.

To test this, we had participants perform trials on three tasks, one that required mentalizing and two others that did not. The trials were intermixed so that participants could not guess which kind of trial was coming next. We also gave participants short rest periods (two to eight seconds) between trials. We examined the extent to which the default network came on during these brief rest periods and how that related to performance on the trial that followed it. Amazingly, participants performed better on the mentalizing trials that came right after a rest period with strong default network activity than they performed on the mentalizing trials that came right after a period of weak default network activity. The same did not hold for the nonmentalizing task trials; the strength of the default network activity right before nonmentalizing trials did not predict performance on those trials. This study provides compelling, albeit preliminary, evidence that default network activity primes us to be social, preparing us to see the world in terms of the mental states of those around us. Thanks to the mentalizing system, we do not see bodies as mere bodies but rather as sentient vessels directed by minds. Evolution could have promoted other systems to come on during breaks, to prepare us to see the world in terms of its mathematical properties or some other nonsocial lens. But evolution made this "choice"—for the brain to reset to thinking socially, and to mute the impact of nonsocial thinking, every chance it gets.

Social Thinking Is for Social Living

We recruit the mentalizing network hundreds of times a day in order to make educated guesses about what is going through someone else's mind. Sometimes, such activity is simply the result of internal musings because we are naturally curious about why people do the things they do. Certainly the mentalizing studies described above give this impression because they involve a detached observer with no connection to the people being observed. However, it is unlikely that we evolved the capacity for mentalizing just so that we could be a fly on the wall. The philosopher and psychologist William James famously wrote, "My thinking is first and last and always for the sake of my doing." This is true for our social thinking too. Often, our success at something is intertwined with how well someone else is doing, or it depends on our interaction with that person. In these cases, keeping track of or predicting the other person's mental state can be the difference between success and failure.

Imagine you and a friend are playing a videogame in which the two of you need to trap an animal in a maze. There are no dead ends so you can't corner the animal by yourself. Instead, you need to coordinate your actions so that you and your friend surround the animal on either end of a path, leaving it no escape. Also imagine that you and your friend are not together but are playing over the Internet, so you can't discuss your strategy. However, you can see your friend's moves, and you have to decide your own next move based on where you think your friend is headed. Neuroscientist Wako Yoshida ran a neuroimaging study on a version of this task, called *Stag Hunt,* and found that the more difficult it was to predict your partner's next move, the more the mentalizing system was re- cruited. Note that although we can use language to explicitly share our intentions in daily life, our primate ancestors could not use lan- guage to facilitate cooperation. Their lack of linguistic skills meant that if larger groups were to coordinate hunting or avoiding preda- tors, a lot of the work had to be done from simple observable cues provided by other members of the hunting party.

We are just as often competing with others as cooperating with them, and in these situations, accurately decoding the goals and intentions of others is all the more important, as others may in- tentionally try to misdirect us. To the untrained eye, card games like poker seem largely luck-based along with a small amount of explicit knowledge about which hands beat others or the likelihood of drawing a particular card to complete a flush or a straight. Any professional poker player will tell you it's nearly all skill. Skill #1 is patience. We all want to play and win every hand. It is more fun to be in the game than to fold and watch from the sidelines. But most dealt hands are bad, and the urge to be in the action will wipe out your chips quickly if unchecked. Winning big requires know- ing when to cut your losses from moment to moment. Skill #2 is bluffing. Can you persuade someone else that you are sitting on a full house when you have nothing but junk cards? Can you get the other person to fold, in which case no one ever finds out that you were bluffing in the first place? Skill #3 is identifying when someone else is or isn't bluffing. When players are matched on skill #1, the game is largely determined by battles between bluffing skills and bluffing detection skills.

The mentalizing arms race can escalate, with each side using countermeasures to outmaneuver the other. In an old episode of *M*A*S*H*, Winchester, the arrogant blue blood doctor disliked throughout the army base, plays a series of poker games with Hawkeye Pierce and others. He cleans them out every time, and it drives them mad (and broke). Toward the end of the show, they realize he has a *tell*, a sign that indicates he is bluffing. They realize that Winchester whistles more loudly whenever his cards are worse than his betting would suggest. By the end of the episode, Winchester has lost everything. On the show, Winchester's role is always to be the butt of the joke; in real life, however, he might have figured out that the others were on to his tell and then used it strategically to his own advantage. He could have changed when and how he whistled to drive the betting up when he had a great hand and to drive others out when he did not. Naturally, the adjustments could go on ad infinitum.

Giorgio Coricelli conducted a study that captured this mentalizing arms race phenomenon. In this study, individuals had to pick a number between 0 and 100 on several different trials. The rules for winning changed from trial to trial, but they always had to do with how the individual's own guess related to the guesses of all the other participants in the study. For instance, in one trial, the rule was that the winner would be the person whose guess was closest to two-thirds (2/3) of the average of all the guesses. This meant each person's guess affected what the right answer was. Someone who was being completely nonstrategic might have generated a random guess between 0 and 100, ignoring the rule altogether. A subject who was slightly more strategic might have thought about all the nonstrategic players, assumed that their average guess would be 50, and thus guess 33 (that is, 50 × 2/3) for himself. A more strategic person might have thought everyone else would be slightly strategic, assume that they would guess 33 on average, and thus guess 22 instead (that is, 33 × 2/3). This strategizing could keep going until one ultimately reached a Nash equilibrium of 0. In other trials, participants guessed a different proportion of the average guess (for example, 1/2 or 3/2). Coricelli computed a measure of *strategic IQ*, which indicated the extent to which the individual made guesses that took into account the possibility of others being strategic. Strategic IQ was highly correlated with activity in the DMPFC, but not at all with activity in the lateral frontoparietal regions commonly associated with nonsocial IQ. Strategic IQ looks a lot like social IQ, and it is linked to the mentalizing system in the brain.

Information DJs

Growing up, I never thought too much about the disc jockeys (DJs) that I would hear on the radio. My British friends in graduate school were obsessed with particular DJs the same way I was obsessed with my favorite band. When I started going to clubs in Los Angeles, a ritual that has long since ceased, I finally understood what made some DJs great. There is an endless amount of music out there in every genre, too much for me to ever sift through. Great DJs spend their time filtering through all of it, and they have the ears and judgment to appreciate which tracks played at certain points in the evening at a particular venue with a specific audience will really get everyone going. While most people listen to music primarily for personal enjoyment, music DJs listen to music to figure out whom they can share it with and how best to do so.

In a sense, the Internet and social media have made *Information DJs* out of all of us. Millions of people post to Facebook and Twitter every day in the hopes that something of interest to them will be of interest to others as well. When I come across the latest research about the social brain or a really cool technology story on *Gizmodo*, I post them to Twitter because I know that many of the people who follow my Twitter feed are interested in these posts. I don't post pictures of my son doing silly things to Twitter because that's the wrong outlet. My family and friends follow my Facebook posts, so off to Facebook those go (with apologies to my Facebook friends who never want to see another picture of someone's kids). Being an Information DJ involves being able to select what to share and knowing one's audience well enough to know how best to share it.

A few years ago, Emily Falk and I became interested in what goes on in our minds when we are first exposed to information that might be relevant to other people. Do we initially take in information in a purely self-interested manner, focused on how the information is useful or enjoyable for us? We wondered if perhaps people are always filtering new information to see how it might be useful or enjoyable to others we might share it with. Being the bearer of good news or the teller of good stories is a great way to become more socially connected.

To examine this, we had people lie in a scanner while we showed them information about possible television pilot ideas (that is, ideas for new shows). We made up these pilots, and we showed people titles, descriptions, and iconic images for each show. After participants got out of the scanner, they had a chance to share their views on which shows should receive further consideration and which should be canned. They were asked to imagine they were Interns working at a television network (for example, NBC) helping to triage the submissions so producers could spend their time considering only the best ideas. Other participants played the part of the Producers, and because they were never shown the original pilot descriptions, they knew only what they heard from the Interns. Finally, we asked the Producers to tell us how excited they would be to pass each show idea on further to, say, network executives.

We were interested in what was happening in the brain of an Intern, the first person seeing the information about the pilot, as it related to whether that Intern would share that idea successfully enough such that the Producer would be excited to pass the idea on even further. When Interns saw an idea that they would later pass on effectively, ensuring that it would spread beyond the Producer, the mentalizing system lit up like a Christmas tree. With few exceptions, the rest of the brain showed very little sensitivity to whether the idea would be spread successfully to the Producer and beyond.

We might have expected reasoning or memory systems to be associated with this effect because committing the idea to memory would seem to help a person communicate better about it later. But instead, we saw the mentalizing system. This suggests that even at the moment we are first taking in new information, part of what we do is consider whom we can share the information with and how we can share it in a compelling way given the individuals we choose to share it with. It's important to note that this effect was not a result of some show ideas being universally liked. Interns had very different rankings of the show ideas, and thus what we were seeing had to do with Interns' ability to communicate their take on a show to the Producers.

We also looked at how the Interns differed from one another. Some of the Interns were better than others at making sure that the Producers came away with the same view of each of the pilots as they had. In other words, some of the Interns were better salespeople. Thus, we looked for a *salesperson effect* in the Interns' neural responses while they were seeing each of the pilots. Only one region of the brain, the TPJ within the mentalizing system, was more active in those participants who were in general better at selling their ideas to others. These findings suggest that, much more than we realize, the mentalizing system is always at work filtering the influx of information we are exposed to each day and selecting for what we should be passing on to others, to help them and to enhance our social connections with them. Once again, we see how mindreading promotes connection.

Practice Doesn't Always Make Perfect

The mentalizing network does something incredibly special to facilitate our dealings with other people. It allows us to peer inside the minds of those around us, take into account their hopes, fears, goals, and intentions, and as a result interact with them much more effectively. It allows us to figure out the psychological characteristics of people we see every day so we can better predict their reactions to novel situations and avoid unnecessary feather ruffling. We use these abilities to achieve cooperatively things that we never could do on our own, as well as to strategically compete with those around us. The mentalizing system allows us to filter our experience to figure out the best information to share with others and how to do it. We would be absolutely lost without our all-purpose mindreading machine.

How effortlessly does our mentalizing system work? Does it work only when we are consciously trying to use it, like a working memory system? Nobody counts backward by 17s unconsciously or accidentally? Or does mentalizing work more like vision, which causes us to see, automatically, whenever our eyes our open? The answer to this question is pretty complex; however, we have good reason to think that although the mentalizing system comes on spontaneously, it does operate like a working memory system, a *social working memory system.* Meghan Meyer and I conducted a study in which we asked participants to perform a working memory task. But instead of holding numbers or letters in their minds as in a nonsocial working memory task, they had to think about several of their friends in terms of how funny, persistent, or anxious they were. Just as in a nonsocial working memory task, the harder the trials, the less likely participants were to get it right and the longer they took when they did. However, unlike nonsocial working memory tasks, for which harder trials turn off the mentalizing system, in this study, harder trials led to greater increases in mentalizing system activity than easier trials. In a follow-up study, we found that performance on a social working memory task was essentially uncorrelated with performance on more traditional working memory tasks, suggesting that it really is a distinct psychological process.

So the mentalizing system appears to require effort to function effectively in most contexts. This matters because humans hate exerting effort and thus may not use their mentalizing system as well as they could in everyday life. If there is a way to avoid exerting effort, we almost always do. If there's a mental shortcut we can take to avoid hard work, that's the route we will take. These shortcuts are called *heuristics*, and we use them all the time to simplify decision making. Heuristic processes evolved because they do well enough in most situations and represent a reasonable trade-off between accuracy and effort. But they can get us into trouble at times.

Heuristic shortcuts are no less prevalent when it comes to social thinking. Even though adults may pass all of the Theory of Mind tests with ease when they know they are being tested, adults do not always fully apply this ability in everyday life. There is a big difference between having the capacity to do something and actually using this capacity unprompted. In daily life, we often use a lower effort heuristic in place of the hard work of accurate mentalizing. We often use our own mind as a proxy for other minds, acting as if what we see, others see, what we believe, others believe, and what we like, others like. Rather than figuring out whether our friend would like a movie based on a careful assessment of movies that the friend has liked and disliked in the past, we often just assume that if we liked it, our friend will too. If you are thinking about *Avatar*, the most successful movie in history, this heuristic won't get you into trouble too often. However, if you are one of the handful of people in the world who, like me, thought *Eyes Wide Shut* was a worthwhile film, it is not a good idea to assume others have this preference as well.

Boaz Keysar, a psychologist from the University of Chicago, created an elegant paradigm called the *director's task* that has been used to demonstrate the limits of mentalizing in adults. Imagine sitting down with another participant at a table; between you is a 4-by-4 grid of shelves (see Figure 5.4). Some of the shelves have a small object on them like a toy car or an apple. Some of the shelves are covered on one side so that you can see what's on that shelf, but your partner can't. In all, there are 16 spots. You can see all of them, but your partner can see what's in only 12 of them. The game you are asked to play involves moving the objects in whatever way your partner ("the director") asks you to (the experimenter has given your partner a script of what moves to ask for). Let's say your partner asks you to move the toy car down one space. That's easy. Apple over two spaces to the right—no problem. But there's a special type of request that gets tricky.

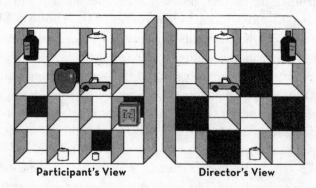

Participant's View   **Director's View**

Figure 5.4 The Director's Task. Left shows the participant's view; the right shows the director's view. Adapted from Keysar, B., et al. (2000). Taking perspective in conversation. *Psychological Science*, 11(1), 32–38.

Notice there are three candles on different shelves within the grid, the smallest of which you can see but your partner cannot.

What should you do when your partner tells you to move the "small candle"? Iroise Dumontheil and Sarah-Jayne Blakemore asked young children, teenagers, and adults to perform this task. When faced with the candle trials, the young children moved the wrong candle almost 80 percent of the time. Typically, the children would move the smallest candle—that is, they would move a candle that their partner could not see and thus could not have been referring to. Such behavior is egocentric because the children appear not to be considering their partner's perspective and instead act as if their partner can see what they see.

Adults do much better than children in this task. And they should, given that their mentalizing ability is much better developed. However, adults do not do nearly as well as we might guess. Most of us would imagine that it might take slightly longer to get the tricky trials right because there is more to consider, but we also believe that we would get these trials right nearly every time. If you know your partner can't see the smallest candle, why would you ever move that candle? However, in Dumontheil and Blakemore's study, the adults made mistakes on the tricky requests 45 percent of the time. Yes, adults have the capacity to mentalize well, but as this study shows, they don't apply this tendency reliably. This is probably because the brain regions that support accurate mentalizing require effort to work well, and we are wired to be mental couch potatoes whenever we can get away with it. We may mentalize a great deal, but that doesn't mean we always do it well or that we can't learn to do it better.

The Miracle of Mentalizing

Although we begin to gain the capacity to appreciate the differing beliefs and perspectives of others in our preschool years, even as adults we continue to use this capacity somewhat inefficiently. Nevertheless, mentalizing is one of the signature achievements of the human mind, one that separates us from all other species. Along with our capacities for language and abstract thinking, mentalizing is the primary reason we live in homes with air-conditioning and communicate over tiny wireless devices. No business, classroom, or friendship can thrive without this miraculous mental process. Mentalizing allows us to imagine not only what other people are thinking or feeling right now but also how they would react to nearly any event in the future. It even allows us to consider how their reactions would change as their development, interests, or circumstances change.

Apple cofounder Steve Jobs suggested that his own view on product design was much like that of Henry Ford, who famously said, "If I'd have asked my customers what they wanted, they would have told me, 'A faster horse.'" The essence of successful inventing, Ford would say, is to figure out what people will want before it exists. Steve Jobs was a master at understanding what we would want better than we could guess ourselves. The iPod was declared dead on arrival when it was first announced in 2001. By 2011, more than 300 million iPods had been sold, not counting the iPhones, iPads, and countless rival devices it inspired. The idea of the iPod may not have been inspiring to most, but Steve Jobs bet the entire future of Apple on his belief that when other people experienced his products, they would love them.

In little ways, every day, we use mindreading to anticipate the desires and worries of the people in our lives and act to make their lives a bit better. When we are lucky, they do the same for us. Our ability to mentalize is the difference between social pain and pleasure being random occurrences and their being destinations that we can navigate toward or away from.

CHAPTER 6

Mirror, Mirror

A friend of mine once joked that if he ever discovered he was living in a counterfeit world like Jim Carrey in the movie *The Truman Show,* his first response would be, "Airplanes! How did I ever get tricked into believing 300-ton metal buses could actually fly?" As impossible as it seems, planes take off and land uneventfully thousands of times a day. Flying is one of the safest modes of travel, only slightly riskier than using Google Earth to see the world. Of course, air travel wasn't always so safe. Flying through the air at hundreds of miles per hour means that any system failure has the potential for catastrophe, or at least it did until airplanes began being built with massive redundancy. Engines, flight controls, and communications equipment are each duplicated within the same aircraft to ensure that if one fails, the plane will still reach its destination safely. Flying is a high-stakes enterprise that is well worth the extra dollars to prevent fatal system failures.

Our ability to understand what is going on in the minds of others may not have the same life-or-death consequences as an airplane crash, but over the course of a lifetime, making sense of the thoughts and intentions of others can be the difference between increased happiness and social connection or escalating loneliness and frustration. It might make sense, then, if evolution selected for brains with redundant systems for making sense of other people.

In this chapter, we examine a second neural system that has been associated with making sense of other people—one that has a radically different architecture from the mentalizing system. Unlike the mentalizing system, this second system is shared by humans and other primates. Various claims have been made about the two systems regarding which is superior for making sense of others. As often happens in science, defenders of each system tend to study their preferred system under conditions that maximize what can be attributed to that system and minimize the apparent contributions of the other system. In reality, the two systems perform different jobs that are most often complementary, each critical to making us the massively social creatures that we are. Both help us make sense of the ordinary actions of others every day. Both are vital in allowing us to empathize with others and experience compassion for their misfortune. And both are aberrant in autism, a condition that leaves individuals unable to understand the minds of others easily and thus less able to form and maintain important social bonds.

Monkey See, Monkey Do

Giacomo Rizzolatti at the University of Parma in Italy specializes in primate neurophysiology. Throughout the 1980s, his lab was focused on examining how individual neurons in macaque monkeys responded when a monkey performed an action. Some neurons in the premotor cortex would respond selectively when the monkey grasped an object with its hand. Other neurons responded when the monkey put an object in its mouth. Some neurons responded to the sight of an object that could be grasped, even if it was not being grasped at the moment, while other neurons did not respond to the sight of the object unless the monkey was acting on the object. In other words, primates have a lot of different neurons responsible for different functions related to performing even the simplest action.

While conducting one of these studies, the researchers noticed something unexpected. Their serendipitous discovery has, to many minds, changed our fundamental understanding of how we came to be such social creatures. Some of the same neurons that responded when, say, a monkey grabbed a peanut with its hand also responded when the monkey watched the scientist grab a peanut. These neurons did not respond to the sight of the peanut alone, in the absence of action directed at the peanut. And they did not respond to the sight of the experimenter pretending to pick up a peanut when there was no peanut there.

These results were startling because neuroscientists had thought of the brain as being divided into different sections for perceiving, thinking, and acting. But in these *mirror neurons*, perception and action were occurring in the same exact neuron. Picking up a peanut and seeing another person picking up a peanut had the same effect on these neurons. Although some psychologists had argued for this kind of perceptual-motor overlap before, it was a revelation to most. The neurons that responded to action weren't supposed to be involved in perception. But Rizzolatti's work suggested these neurons might be.

The excitement over the discovery of mirror neurons grew in such a way that they quickly became the solution du jour for many of the hardest problems within psychology. Championing this perspective, renowned neuroscientist V. S. Ramachandran wrote that mirror neurons are "the single most important . . . story of the decade" and argued that "mirror neurons will do for psychology what DNA did for biology: they will provide a unifying framework and help explain a host of mental abilities that hitherto remained mysterious and inaccessible to experiments." Indeed, as he predicted, a long list of mental phenomena have been attributed to mirror neurons since their discovery, including our capacity for language, culture, imitation, mindreading, and empathy. That's pretty heady stuff—a single neuron explaining all of these miracles of humanity.

Exciting new discoveries in science often go through a Hegelian three-step waltz, starting with the promise that the discovery will account for 100 percent of the unexplained phenomena (phase I: thesis), followed by loss of faith that the discovery explains much of anything (phase II: antithesis), and eventually settling into a realistic appreciation of what the discovery does and does not contribute (phase III: synthesis). Mirror neurons are probably somewhere between phases I and II today; they are still promoted as something of a cure-all in some corners, but they also have a growing chorus of vocal detractors. I personally began more in this second camp, but I think I will end up most comfortable when phase III is in full swing. Ultimately, I think mirror neurons do two very important things. First, mirror neurons play an important role in our ability to imitate others. Second, mirror neurons do something essential that allows mindreading to occur, but I believe it is more of a behind-the-scenes role than is generally understood.

Imitation

The human brain reached its modern size around 200,000 years ago, and yet there is little evidence of advanced culture (for example, complex tools, language, religion, or art) prior to 50,000 years ago. It has been suggested that some minor genetic change around that time served to push us over a tipping point, creating a cascade of self-reinforcing cultural development. Some have argued that this genetic change enhanced our working memory system, allowing us to keep more abstract ideas in mind at the same time. Ramachandran has countered that a genetic change affecting mirror neurons may have accelerated our development, characterizing this change as "the driving force behind the 'great leap forward' in human evolution."

Our cultural development of skills and habits depends on our capacity for imitation. Given that mirror neurons respond during both action and perception of an action, promoting the fine-tuning of one's own actions in light of what is seen, they seem like an ideal mechanism to support imitation and imitation-based learning. Especially in a prelinguistic society, the ability to learn through imitation is likely to be the chief method of spreading innovation from person to person, and from generation to generation. Minor innovations to any procedure such as hunting or creating shelter can be passed on to others, who can add further innovation in a beneficent spiral. Were mirror neurons the original social media—a way to share what we knew before we were able to say it out loud, send tweets, or post status updates to the cloud? Increasingly, the answer looks to be yes; mirror neurons seem to play a key role in imitation.

In 1999, my colleague Marco Iacoboni published the first evidence regarding the presence of a mirror neuron system in humans. Rather than focusing on action and observation, like the prior work in monkeys, Iacoboni focused on observation and imitation. Individuals in his study were shown visual displays of finger movements while being brain-scanned, and they were asked to either watch the images or imitate them. Iacoboni found that regions similar to those Rizzolatti had seen in monkeys were active during both observation and imitation in humans. This suggests that these regions in the lateral frontal and parietal areas (see Figure 6.1) have mirror properties similar to those observed in mirror *neurons* in monkeys. Because fMRI does not look at individual neuronal activity, studies like this one cannot claim to have found mirror neurons per se in humans. Thus these regions, specifically the premotor cortex in the frontal lobe along with the anterior intraparietal sulcus and the inferior parietal lobule, are often referred to as the *mirror system*, not the "mirror neuron system," in humans. Note that while the mirror system and the working memory system both reside in the lateral

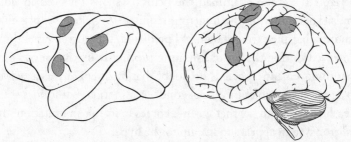

Figure 6.1 The Mirror System in Macaques (left) and Humans (right)

frontal and parietal cortices, they are actually in different locations within these regions.

Although this first imaging study suggested a role for the mirror system in imitation, two additional kinds of evidence are necessary to make the case. Iacoboni's next step was to test whether imitation was affected when the mirror system was temporarily impaired. His group used *transcranial magnetic stimulation* (TMS), a technique that directs an electromagnetic field at a particular spot in the cortex and temporarily "frazzles" the neurons in that area such that the region is essentially taken offline. It sounds like a scary technique, but when done properly with healthy individuals, it is a safe and temporary procedure. In this study, individuals were asked to imitate sequences of button presses as TMS was applied. When TMS was applied to a mirror system region, participants made more errors in their attempts to imitate the other person. But when TMS was focused on a nonmirror region, there was no increase in error rates, suggesting that the mirror system plays a causal key role in imitation.

These studies demonstrated that the mirror system is involved in rudimentary forms of imitation when the to-be-imitated behavior is not novel. Adults are already experts at tapping their fingers before they arrive for these studies. To test whether or not mirror neurons support imitation-based *learning* (that is, spreading new ways of doing things), evidence was needed showing that the mirror system is involved in acquiring new behaviors through imitation. Rizzolatti's group examined the neural systems involved in nonmusicians imitating the fingering required to make a set of guitar chords that they were shown. As predicted, the mirror system was involved during the act of imitating previously unknown complex hand movements. There is no doubt that other cognitive capacities play important roles in various kinds of imitation, but it does seem reasonably safe to say that when it comes to its role in imitation, the mirror system appears to live up to the hype.

Mindreading Mirrors?

A second major claim made by mirror neuron researchers is that the mirror system is responsible for understanding the minds of others. This is the claim that interests us the most as we try to understand the social mind. Just as President Clinton's impeachment fate depended on a verbal dance around the precise meaning of words like *sex* and *is*, whether or not we end up believing that mirror neurons help with mindreading and interpreting the intentions of others will depend on what we mean by words like *mindreading*, *goals*, and *intentions*. To understand the relation of the mirror system to mindreading, we have to retrace our steps a bit to a philosophical debate over how we know the minds of others.

In the early 1980s, developmental psychologists were all abuzz with the notion of *Theory of Mind*—that we have a theory that other people have minds, and with this theory we can logically infer their thoughts, beliefs, and desires in countless situations. Various philosophers like Daniel Dennett and Stephen Stich, one of my undergraduate mentors, were very supportive of this account of how we make predictions about the behaviors of others. For a few years, Theory of Mind was the only game in town. But in 1986, philosopher Robert Gordon suggested an alternative for how we understand the minds of others.

Gordon's key insight was that there are multiple possible ways by which we can predict another person's intentions in a given situation. One route was the one associated with Theory of Mind. Given our theory of how minds in general operate, we could use propositional *if-then* statements to logically figure out what a person's intentions might be. For instance, if we know that someone has not eaten in eight hours, then we can infer he is probably hungry, and if we can infer that he is hungry, then we can also infer that he currently possesses the intention of finding something to eat.

The second route involves imagining what it would be like to be in that situation ourselves and to use our own natural reactions to this simulated experience as a guide to how another person is likely to think, feel, and act. If I want to understand what someone is experiencing after getting dumped by a romantic partner via text message, I can try to mentally re-create the scene, imagining myself as the protagonist. What reactions do I see myself having as this scene unfolds in my mind? They may help me understand how someone else might react.

Often these two routes lead us to reach the same conclusion but through different processes. In the one case, I am thinking logically *about* the situation and how anyone would likely respond to it. In the second case, I imagine myself *in* the situation in order to find out what my own reaction would be. In the first case, my accuracy depends on the quality of my logic, and how similar the individual's mind is to the typical mind I am reasoning about. In the second case, my accuracy depends on the quality of my re-creation of the scene and how similar the target's mind is to my own. Gordon's account of the second route has been referred to as *Simulation theory*; and there is little question that at least in some cases we project from our own experience to that of others.

The key question here is whether mirror neurons have anything to do with either of these accounts. Vittorio Gallese, one of the original discoverers of mirror neurons, has argued that mirror neurons are the neural implementation of Simulation theory. Moreover, he has argued that this is *the* way we come to know the minds of others in normal circumstances, suggesting that "the fundamental mechanism that allows us a direct experiential grasp of the mind of others is not conceptual reasoning but direct simulation of the observed events through the mirror mechanism."

When I think of the kind of mental simulation that Gordon and others have generally described, it sounds like hard work. They suggest mental simulations are analogous to building wind tunnels to test airplane wings in simulated flying conditions or building complex computer simulations with countless variables to do the same digitally. When they discuss social simulations, they frequently talk about mentally constructing all the relevant aspects of a situation before running themselves through the simulation to see how they would react. That sounds like a lot of work. But if as Gallese suggests, the mere perception of another person allows us to intuitively and automatically understand their experience, that could make the mirror system a much more plausible implementation of Simulation theory.

Here's how Gallese's argument works. The fact that your "reaching-for" neurons are active when you see someone else "reaching for" something literally means that neurons in your brain are matching the neural state of the person you are observing. When you see a person you are looking at "reaching for" a cup, both your and her "reaching-for" neurons are active. Gallese and others characterize this as *motor resonance* between you and the other person. If you are experiencing the same motor state as another person, your brain is in essence simulating key aspects of the other person's brain, allowing you to automatically understand the mental state of that person as it relates to her action or activity. My brain is mirroring your brain, and thus by simply knowing the state that I am in, I know your mind as well. In other words, mirror neurons would seem to provide us with an almost magical mindreading device that operates automatically whether we are trying to understand the other person or not.

Cracks in the Mirror

But a growing chorus of critics suggests that the mirror neuron camp has not done enough to prove that the mirror system in humans is central to mindreading. Still other critics think that enough research *has* been done and that the conclusion is clear: the mirror system *is not* central to mindreading. These critics are doing conceptual and empirical work to ensure that our ultimate understanding of the mirror system and its contributions is correct. It's scientific democracy in action.

Mirror neuron researchers have argued that one of the key qualities of these neurons, as they relate to mindreading, is that they are sensitive to the abstract meaning of other people's actions. Imagine you saw someone opening a peanut shell. There's the raw visual information associated with seeing that action. There are also the sounds associated with it. But whether we see or hear the action, our mindreading is focused on the meaning—someone wants to get inside the shell so the peanut can be eaten. On the one hand, if a neuron responds to the sight of the action but not to the sound, or if it responds to the sound but not to the sight, then it is only mirroring something sensory (that is, at the level of the senses). On the other hand, if a neuron is responding to the meaning of another's action, it should not matter whether we see or hear the action. In 2002, Rizzolatti's team found mirror neurons that fit the bill. A subset of mirror neurons responded to both the sound and the sight of actions, suggesting that the increased activity in these neurons could be responding to the meaning of the action, not just to its appearance or sound.

Greg Hickok, one of the most vocal critics of the mirror neuron camp, highlighted an important limitation of the sight-sound study. The researchers had started by identifying neurons that showed the standard mirror neuron properties (that is, neurons that responded during the visual observation and performance of an action), and then they tested those same neurons to see which responded to the sound of the actions as well. Of the neurons initially identified as mirror neurons, only 15 percent also showed the response to the sound of the actions. This means that 85 percent of the mirror neurons responded only to the visual characteristics of an action. As such, they could not be representing its meaning. On the one hand, yes, some mirror neurons in macaque monkeys do seem to respond to the meaning of an action and not just to its appearance. On the other hand, the vast majority of mirror neurons don't do this. This 1-to-5 ratio is important because in an fMRI scan, we aren't looking at individual neurons. Rather, we see the summed effects across large populations of neurons. When the results of fMRI studies are used to claim that the mirror system represents the meanings of actions, there is no way to be sure that it is the meaning-representing mirror neurons that are driving the effects. This analysis does not mean that mirror neurons cannot support mindreading in humans, but it does suggest that it will be very difficult to assess this ability with fMRI.

A second finding that is used to defend the notion that mirror neurons represent the abstract meaning of an action, rather than just the sensory aspects, is that the mirror neurons respond to actions that involve objects that cannot be seen. Rizzolatti's group showed monkeys an object but then placed a dividing wall between the object and the monkey so that the object could no longer be seen. The same mirror neurons that would respond when the monkey saw an experimenter reach for the visible object also turned on when the experimenter reached for the hidden object. Rizzolatti argued that if the mirror neurons were responding only to the visual properties of the action, they would not respond to someone's reaching for a hidden object.

But Hickok points out that this argument is flawed. The monkeys could be using working memory to hold the image of the hidden object in mind. Humans are certainly capable of seeing an object and then continuing to visualize the object in their mind after it can no longer be seen. Perhaps the monkeys are doing the same. Mirror neurons may be responding not to the meaning of the action but rather to a visual representation held in our brains.

Cecilia Heyes, a psychologist at Oxford University, makes a very different argument against mindreading mirror neurons. She suggests that the purpose of mirror neurons cannot be motor resonance–induced understanding of others because these neurons are not intrinsically wired for mapping observed actions or meaning from performed actions. Instead, Heyes suggests, the reason the "reaching-for" neurons are activated both when I perform and when I observe a "reaching-for" action is past experience, rather than an intrinsic mirroring function. Heyes believes mirror neurons are really just motor neurons that become conditioned over time to respond to the sight of our own actions as we perform them (the neurons then generalize to seeing others perform the same actions). Given that sea slugs can be conditioned to learn that one event is associated with another event, conditioning does not imply any sort of meaningful understanding. I have seen my own hand reach for my spoon thousands of times since infancy, so whether mirror neurons are specifically designed for motor resonance or merely conditioned to link the action and the sight of the action, both Heyes and the mirror neuron camp agree that when I see someone else use a spoon, my "spoon-reaching" mirror neurons will turn on. Only in special cases would Heyes and the mirror neuron camp predict different outcomes.

In order to test her account, Heyes designed a clever *counter-mirroring* procedure during which the performance of an action was associated with a *different* visual action. If I am instructed to move my foot whenever I see a hand moving, a true motor resonance mechanism should be insensitive to this behavior. Yet Heyes found that when individuals learn to respond with a behavior different from the one seen, the mirror system is activated just as it is during direct imitation. This outcome suggests that while the mirror system frequently responds to the sight and performance of the same action, this response is not intrinsic to its functioning; it can learn to respond to the sight of one action and the performance of a different action. If mirror neurons link my foot movement to your hand movement, it's hard to see how this response constitutes motor resonance or promotes mindreading.

Another study examined how the mirror system responds when individuals perform a complementary behavior instead of imitating an observed behavior. Imagine that there are two objects on a table: a small object that you pick up by pinching your thumb and first finger together (for example, a sugar cube) and a larger cylinder that you pick up using a cup-like grip (for example, a can of soup). In some trials, individuals were told to imitate the motion and grip used by the other person shown in a video. For other trials, individuals were told instead to complement the other person by preparing to grab the other object using the other grip. Here, if your partner started to pinch to pick up the sugar cube, you would make a cup-like grip to pick up the can of soup. The mirror system was more active during the complementary action trials than it was during the imitative action trials. There is no reason why a motor resonance mechanism would be more active as we engage in a behavior different from the one seen. Along with the previous study, this suggests that the mirror system is not designed solely for the purpose of matching our internal state to that of another person.

What Are Your Intentions?

There have been dozens of mirror system studies of mindreading inspired by Simulation theory and dozens of mentalizing studies inspired by the Theory of Mind accounts. Both research programs have focused on how one person understands the mental states of another person. One might think that whatever one's theory going in, everyone would be looking at the same brain, so the results of these studies would have to converge and tell one story. Although the studies were motivated by different theories, they still should have ended up in the same place because the brain does not care about our theories. It should just show what's true.

Yet despite the numerous fMRI studies from both the mirror and the mentalizing camps, their results almost never come together. From looking at the brain scans of different studies, you wouldn't guess that all of these researchers have been studying mindreading. Not only have the two camps consistently produced findings that do not overlap anatomically, but the regions of the brain observed by each group tend to be inversely correlated with one another. If we look at brains at rest, the more activity people produce in the mentalizing system, the less they produce in the mirror system. These were not just opposing theoretical camps. They were studying what appeared to be opposing neural systems. Yet both systems are supposedly end-to-end solutions for mindreading in which the other system plays no role.

There are two reasons why each camp saw only their preferred regions active when they were ostensibly studying the same thing. First, the two camps study mindreading in very different ways. The mentalizing camp tends to use verbal materials and cartoonish drawings. In other words, the materials are pretty abstract. If the mirror system is activated by seeing real action, it is understandable that mentalizing studies not showing real action would fail to activate mirror regions and overlook their contributions to mindreading. On the flip side, a major strength of mentalizing studies is that they manipulate whether a person is trying to understand the mind of another person. After reading a paragraph that implies an individual's mental state, participants in mentalizing studies are often asked questions about the protagonist's beliefs, motives, and personality—questions that can be answered correctly only if the proper mentalistic inferences have been drawn. Mirror system studies never pose these mental state questions to participants, perhaps because when you see only a disembodied arm, questions about beliefs and personality do not make much sense. Thus, mirror system studies minimize the involvement of the mentalizing system.

The second big issue, the Clintonian verbal hairsplitting issue, concerns the meaning of words like *goal* and *intention*. Let's say you see a friend drinking a glass of single malt scotch at 8 a.m. You ask him why. If he answers, "In order to have a drink," strictly speaking, he is answering your question, providing you with a goal ("to have a drink"). But his answer is entirely unsatisfying. It is obvious that he is taking a drink because he wants to take a drink. What you really want to know is what special motivation led him to the unusual goal of wanting to have an alcoholic drink at this hour of the morning. The responses "to have a drink" and "to drown my sorrows because I lost my job" are technically both answers to the question, but they highlight how the word *goal* can have different meanings.

In the 1980s, Robin Vallacher and Daniel Wegner systematically investigated these distinctions. They conducted a series of studies highlighting how we can understand the same action in different but equally accurate ways. I can describe my current behavior at my computer keyboard as "moving all of my fingers slightly up and down," as "typing," as "writing a book," or even as "trying to share what I have learned with others." Given the order in which I've presented these options, we can see that they form a hierarchy, with the first answers giving lower-level descriptions of specific motor behaviors and the latter ones describing higher-level long-term goals that have greater meaning. No one comes to the end of their life and says, "I wish I had moved my fingers up and down more," but we can imagine someone saying, "I wish I had taken the time to share more of what I learned." We live in a world of meaningful actions that can be described at both high and low levels, but we typically focus on one level at a time, depending on what we are interested in. New typists focus on which fingers they are moving to find particular letters, while experienced typists are more likely to focus on the thoughts they are trying to convey.

Among the biggest differences between mirror neuron researchers and mentalizing researchers are the kinds of goals they are interested in explaining. Mirror neuron folks are focused on how we understand the lower-level motor intentions of others ("He is flicking the light switch because he wants the light to come on"), whereas the mentalizing folks are more interested in higher-level intentions ("He is turning on the light because he wants to study for an exam"). The intentions of the other person are described in both cases, but I think it's fair to say that in everyday life we are usually interested more in the second kind of goal than in the first.

The motor resonance account is well positioned to explain how we understand low-level motor intentions. I see you flicking the light switch, which activates the "light switch–flicking" mirror neurons in me. But these same neurons are poorly positioned to explain another person's high-level reasons for wanting the light on. There are countless reasons why I might want the light on, and many of these are belief based ("I heard a noise downstairs in the middle of the night so I am turning on a light to see if anyone is there" or "I woke up with a great idea for a story so I want to turn on the light in order to write it down"). We don't turn on the light differently in each of these instances, so there is no way for motor resonance to clue us in to which of these higher-level intentions is present. The mentalizing system of the brain is ultimately needed to figure out the higher-level intentions of others. The question is, what role, if any, does the mirror system play in allowing us to understand the higher-level motives of those around us?

How, What, and Why

When we look at another person's behavior, there are three questions that we might be interested in answering. These three questions correspond to different levels of analysis in the Vallacher and Wegner studies. The first, obvious question is *what* someone is doing, which we answer in the most generic action language possible: "She's crossing the street," "He is typing," "The cat is eating my leftovers." We interpret actions this way so regularly that we don't even realize we are doing it unless someone does something really out of the ordinary ("Is he scaling the side of a building?"). Depending on our goals, we might then ask ourselves one of two follow-up questions. If we have any further interest in the person, we most likely want to know *why* the person is doing *what* they are doing: "She is crossing the street *in order to* get to work," or "He is typing *in order to* finish his final paper." Sometimes, though, our interest is less in the person and is instead focused on the behavior itself. We may want to learn *how to perform* the same behavior, such as when a student taking guitar lessons watches his instructor in order to figure out *how* the teacher is doing *what* he is doing.

Bob Spunt and I have conducted a series of studies to examine how the mirror and mentalizing systems contribute to our figuring out the how, what, and why of other people's behavior. Do different systems in the brain handle each of these questions? We intuitively expect different parts of the brain to handle seeing and hearing because the experiences of seeing and hearing are so fundamentally different from one another. I'm not so sure that looking at the same behavior in others and asking *how, what,* and *why* seem different enough to involve distinct regions of the brain. But this is why we actually run the studies.

In undertaking the studies, our thinking was that if you see a behavior (for example, a woman recycling a bottle) and ask the question "Why is she doing that?," your answer will likely involve a high-level meaningful answer that requires the mentalizing system of the brain (for example, "She is a conscientious person," "She wants to protect the environment," or "She wants to impress a guy who she knows recycles"). However, if you simply want to imitate the behavior and ask "How is she doing that?," this response would involve low-level answers that do not require the mentalizing system but would instead rely on the mirror system of the brain that represents motor movements (for example, "She puts the glass and plastics in the blue bin"). This is exactly what we have seen in multiple *why-how* studies. Whether participants are watching everyday actions or watching someone experience strong emotions, asking *why* recruits or draws on the mentalizing system, while asking *how* recruits the mirror system.

We also wanted to find out what the brain was doing when a

person answers the *what* question. With the *why* and *how* questions, we just instructed people to answer those questions. But given that people answer the *what* question in everyday life without stopping to think about it most of the time, we thought that prompting people to stop and think about it might not actually capture the natural process. The way we worked around this was by manipulating whether or not participants needed to spontaneously answer the *what* question before they could answer the *why* or *how* question they were actually given. Sometimes we showed participants a video of a person performing a task (for example, a woman highlighting passages in a textbook), and other times we replaced the video with a description of the action (for example, "She is highlighting passages in a textbook"). Whether answering the *why* or *how* question, the first thing the participant needs to do when the action is shown visually is to figure out *what* is happening. In contrast, when the action is described in words the *what* is already characterized in the description of the action—it is literally the answer to the *what* question. By comparing the visual and verbal presentations of the actions, we can identify brain regions that support the implicit *what* decoding process.

When we looked at the brain to see what regions were more active during visual than during verbal presentations, we saw two things. First, we saw lots of activity in the back of the brain in the visual cortex. This was expected, as video clips contain far more visual information than text. The other thing we saw was increased activity in the mirror system. Whether the explicit task involved asking *why* or *how*, there was increased mirror system activity when the action was presented visually. In fact, this was still the case when participants were distracted by being asked to recite a seven-digit number while performing the *why-how* task. Distracting people in this way is a common technique for identifying processes that are so automatic that they still take place despite participants being distracted. The fact that even with this distraction, the mirror system responded to the visual action suggests that its decoding of *what* is happening is pretty automatic. In contrast, the mentalizing system was significantly less active when people were under cognitive load, suggesting that this system does not do well when people are distracted.

Making the Social World Possible

The *why-how* studies tell us a great deal about what the mirror system does and does not do with respect to mindreading. The mirror system does not generate high-level mindreading on its own. It does not invoke personality or motives to explain why someone would engage in a particular behavior like drinking a glass of single malt scotch at 8 a.m. It is the mentalizing system that is critical in generating satisfying answers when we want to know why someone is doing something.

However, the mirror system does something that is an essential precursor to mentalizing in most everyday contexts. Being able to see a series of body movements as a coherent coordinated action that can be characterized in a few words is a remarkable achievement of the brain. William James famously noted how impressive it is that we see an orderly world of objects rather than a "blooming, buzzing confusion." Given that nothing in the world unambiguously tells us where one object ends and another begins ("Is that a table with a separate mug sitting on top, or is that a table with a mug-shaped protrusion?"), it is striking how easily we do this. Seeing the world of sentient beings in terms of their actions instead of their movements is equally impressive.

Every movement we see could be described with a nearly endless array of motion parameters (angles, direction, torque, acceleration), but these would be impossible to consciously comprehend together and wouldn't tell us anything about the mind behind the movements. It is only by synthesizing the complexity of movement into the simplicity of an action that any psychological analysis of another's goals, intentions, desires, and fears can begin. Movements are not psychological and imply no specific meaning (for example, moving fingers up and down). In contrast, actions are psychological (for example, typing), and although they do not provide high-level meanings in and of themselves, they suggest there are meanings and motives hiding behind them, waiting to be discovered. The ability to identify *what* someone is doing is the first step toward being able to understand *why*. In essence, the mirror system provides the premises that the mentalizing system can then logically operate in in order to answer the *why* question. Thanks to the mirror system, we live in a world of actions, not movements, which allows us to live in a world of meaning.

In the end, the mirror system is what allows us to experience the world as social, full of the psychologically infused behavior of others. Although our mentalizing system can mindread from written sentences without the aid of the mirror system, mindreading based on words is a recent event in our evolutionary history. In the course of ordinary life and certainly in the development of pre-linguistic children, the mirror system is constantly doing the work of preparing the brain for mindreading. The mirror system chops the world of living movement into pieces, and it repackages them into the psychological elements that the mentalizing system can work from. This process is understated—like a chief of staff who mostly works in the background but makes the president's work possible. Primates have long had the mirror system, but only humans appear to have an advanced mentalizing system. Primates live in a world of *what* others are doing, but only humans live in a world of *why*, with the rich meanings and interpretations to explain the actions of those around us.

CHAPTER 7

Peaks and Valleys

In 1992, I graduated from college on top of the world, with great friends, admission to an excellent PhD program, and a three-year romantic relationship that was going strong. Just months later, I was living alone in a dorm room wondering how everything had gone so wrong so fast. The transition to graduate school was not an easy one for me. I was waiting for my advisors to figure out that I was an admissions error—all the other graduate students seemed much smarter and more productive than I was. I had no real friends to speak of yet in grad school; I avoided going to the dining hall for weeks at a time because I didn't fit in with the other graduate students in my dorm. To top it off, my relationship was on the rocks, and I was broke. It was easily one of the most depressing periods of my life. I coped, if one can call it that, by watching hour after hour of reruns and infomercials on television. *Star Trek Next Generation* marathon until 4 a.m.? I'm there. The George Foreman Grill infomercial again? Keep watching.

It was on one of these sad, lonely nights that a half-hour program came on about donating money to improve the life of a young child in Africa, who would otherwise starve to death or die of preventable diseases. You have probably seen countless versions of these, and I had too. But for some reason, that night, I found myself nearly in tears calling to make a donation in the middle of the night. As unhappy and broke as I was, something about the images I saw moved me to try to do something potentially helpful for a stranger half a world away. The suffering of those children momentarily broke me out of my ongoing self-pity about my life and allowed me to feel empathy for people who clearly had it far worse than I did.

On the face of it, my behavior was irrational. I needed more money, not less. I had never met and would never meet the people I was supposedly helping, and they were never going to thank or repay me. I never told anyone about making the donation. I also don't remember feeling anything pleasurable or rewarding about the experience or having thoughts about what a good person I was being. Given that I did it only once and did not renew the donation the next year, I don't take this episode as much evidence for my virtue. When I go back to that moment, I can only say I felt compelled to do it. My feelings of empathy compelled me to right that wrong in whatever small way that I could.

I Feel Your Pain

The word *empathy* was introduced into the English language just over a century ago as a translation of the German word *einfühlung*, which means "feeling into." *Einfühlung* was used in nineteenth-century aesthetic philosophy to describe our capacity to mentally get inside works of art and even nature itself, to have something like a first-person experience from the object's perspective. *Empathy* still means something like "feeling into," but it almost always refers to our connecting with another person's experience, rather than "getting inside" an object.

We have already discussed how helping others produces feelings of social reward as a consequence. However, empathy is a more complex process that serves to get us ready to help others. Empathy is a front-end process that motivates us, whereas the social rewards are the back-end consequences.

There are at least three kinds of psychological processes that come together to produce the empathic state: *mindreading, affect matching,* and *empathic motivation.* Depending on the situation, either our mirror or our mentalizing system provides the inputs that set the empathic state in motion. As we discussed in Chapter 6, the mirror system allows us to understand observed actions as psychological events and may similarly allow us to understand emotional events when the visual scene provides a direct interpretation of what is going on. One way we know the mirror system is sensitive to emotional cues from others is that we tend to literally mirror them, producing motor responses consistent with the other person's experience.

For instance, when you see someone else getting a shock applied to her forearm, you are likely to clench your own fist and wince at the pain. In one study, individuals watched others receive shocks to their hands or feet. The observers produced electrical responses in their own hands or feet, mirroring what they saw. When they saw someone's hand shocked, their brain sent a signal to their own hand, but when they saw someone's foot shocked, their brain sent a signal there instead. Similarly, when we see another person's emotional expression, muscles in our own faces immediately mimic the expression in subtle ways. And if a person is unable to mimic those facial expressions because of recent Botox injections that actually paralyze the expressive muscles in the face, that person will actually be worse at recognizing emotions in others.

Thus, the imitative responses we have when we see other people's emotions actually help us to instantly understand those experiences. Given that the mirror system is involved in understanding the psychological meaning of other people's movements and is involved in imitating them, it is not surprising that the mirror system has been implicated in studies of empathy and emotional imitation.

Sometimes seeing someone's emotional expression doesn't allow us to fully understand a person's experience and empathize with it. Imagine someone walks toward you beaming with a giant smile. Your mirror system may help you intuitively understand *what* that person is feeling, but without knowing *why* the person is feeling that way, it is difficult to empathize and share in the joy. Is that smile the result of getting a good grade on an exam or getting engaged? In many situations, trying to understand why someone is experiencing a particular emotion ultimately depends on the mentalizing system. Because of the flexibility of the mentalizing system, humans are capable of empathizing with events they have not observed or experienced themselves. Your mother might tell you that your uncle didn't get the promotion he was hoping for. The mentalizing system is likely the key to understanding your uncle's experience or even the "experience" of a character in a novel. Indeed, those who read fiction tend to have stronger mentalizing abilities, suggesting that engaging with fictional minds may strengthen this system.

Whether our understanding of the other person's experience is coming through the mirror system, the mentalizing system, or both, this is still only one piece of the puzzle, not the full empathic state. I can imitate without feeling. I can understand without feeling. I can understand the dread a dictator must feel as his hold on power is crumbling, but my understanding is more likely to promote schadenfreude—the pleasure that results from seeing the misfortune of another—rather than empathy. Empathy really only occurs when the information our brains have gathered via the mirror or mentalizing system leads to affect matching and empathic motivation.

Researchers examining the neuroscience of empathy have spent a lot of effort studying the affect matching component of empathy. Indeed, it was the most famous neuroscience study of empathy that put affect matching on the map. Tania Singer, then working at the University College of London, ran an empathy study that closely mimicked the original mirror neuron studies run in monkeys. But instead of looking at the neural responses when someone reaches for a peanut or sees someone else do it, Singer examined the brains of people receiving an electric shock or watching someone else receive it. She asked women to lie in the scanner while their boyfriends sat nearby outside the scanner. On different trials of the study, either the woman inside the scanner or her boyfriend outside the scanner was shocked using electrodes that were attached to their arms. The woman could see her boyfriend's arm as shocks were delivered to him. As in mirror neuron studies, Singer and her colleagues looked for regions of the brain that were activated both when the women received painful shocks and when they saw their significant others being shocked in the same manner.

Singer found that the women activated the *pain distress network* in the dorsal anterior cingulate cortex (dACC) and the anterior insula regions of the brain (see Figure 3.2) regardless of who was receiving painful stimulation. These women could say to their boyfriends with a straight face, "I feel your pain." People have made claims of empathy throughout history, but it has never been clear whether these have been more than social gestures. Singer demonstrated that it may literally be painful to watch a loved one feel pain. And not just metaphorically painful, but painful in a way similar to feeling one's own physical pain.

Our Better Angels

Affect matching is an extraordinary capacity that can shake our very being, but on its own, it is still not the full state of empathy. The neuroscience of empathy has focused mostly on the neural bases of affect matching for pain distress. This is a natural place to start, given that we have such strong reactions to the pain of others. However, all the focus on empathy for pain has brought with it some limitations as well.

First, affect matching when we see someone in pain does not always lead to prosocial empathic responses. At the opening of the chapter, I described how I was moved to donate money to help disadvantaged children in Africa. What I didn't mention were the countless times that I had previously seen similar ads and changed the channel because it was too distressing. Even though I was affect matching (that is, their distress caused me to feel distress), my focus was on how to alleviate my own distress, rather than theirs. In other words, affect matching can sometimes lead to avoidance behavior rather than the empathic motivation to help. The same distress network in the brain would likely be activated whether I was focusing on my distress or theirs. It is generally agreed that empathy occurs only when there is an appropriate emotional response (that is, affect matching) combined with a sustained focus on the other person's situation, rather than our own. So there is more to empathy than just affect matching.

There's a second limitation to the near exclusive focus on empathy for pain within neuroimaging research. The neural systems responsible for affect matching should vary as a function of what kind of affect is being matched. Given that nearly all of the follow-ups to Singer's seminal work have focused on empathy for physical pain, one could easily review this literature and come to the conclusion that the dACC and the anterior insula are the central mechanisms supporting empathy in general. Is this really the case, or are these regions showing up because they are involved in pain and most of the studies focus on pain?

Finally, almost none of the studies that have been done have linked neural responses during the empathic state to actual helping behavior. One of the purposes of feeling empathy seems to be to motivate us to help others in distress, yet it's unclear how the brain converts our understanding and affect matching processes into the empathic motivation to help.

Sylvia Morelli, Lian Rameson, and I ran an fMRI study that we hoped would capture all three components of empathy: understanding, affect matching, and empathic motivation. First, we varied whether context was needed in order for the key event to be understood. Some of the time, we showed participants a picture of someone experiencing pain that showed all that they needed to know to immediately grasp the event (for example, a hand being slammed in a car door). On other trials, the picture showed someone looking happy or anxious and required context to be understood (for example, "This person is waiting to get his medical test results"). The first kind of event recruited the mirror system; the context-dependent events recruited the mentalizing system. In addition, though the happy and anxious events both included context, they required different kinds of affect matching because they involved different kinds of emotional events. The anxious events and the pain events both activated the pain distress network, but the happy events instead activated a region of the ventromedial prefrontal cortex that is commonly recruited during reward tasks.

Perhaps most important, we searched for brain regions that were commonly activated across all three kinds of empathy events we had included (pain, anxiety, and happiness). Our thinking was that while understanding and affect matching differ depending on the content driving one's empathy, the ultimate empathic motivation to help should be the end result in each case. There was only one brain region that was activated during each type of empathy event: the septal area (see Figure 7.1). In addition to showing up in each condition, the septal area appeared to be a marker of empathic motivation. The subjects that we put in the scanner filled out a survey each day for two weeks about things they did and did not experience each day. Among the questions we asked each day was whether they had helped someone else during that day. By averaging across the two weeks of responses, we had a measure of which people tended to help others more frequently in their daily life. Those who showed more activity in the septal area when they performed our empathy task in the scanner were the same people who tended to help other people more often outside the scanner. This is consistent with the notion that the septal area takes the converging inputs from other brain regions involved in empathy and converts them to the urge to be helpful. What was different on the night that I donated to help a child in Africa, in contrast to the countless other times when I changed the channel? Probably septal area activity.

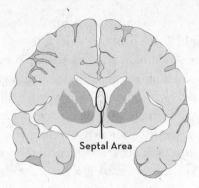

Figure 7.1 The Septal Area

The Septal Area

If I had to bet on which brain region is most ignored by the field of social neuroscience today but will be the hot area of study in the next ten years, the septal area would be it. This structure has become disproportionately larger across primate evolution, and it has direct connections to the dorsomedial prefrontal cortex (DMPFC)—the CEO of the brain's mentalizing system. The vast majority of the work on the septal area has been done in rodents, rather than humans, in part because it is such a tiny region that it is hard to identify with fMRI. The downside to studying rodents is that we can't measure their experiences or even verify that they have them. The upside is that more invasive studies can be conducted to examine how individual septal neurons respond or how surgical removal of the septal area alters behavior.

Animal research provides clues to what the septal area does, but these clues seem to lead us down different paths. Some of the earliest research on the septal area focused on pleasure and reward. Although the ventral striatum has far more often been identified with reward, its neural neighbor, the septal area, was actually the first brain region identified with reward processes. The brain's reward system was first discovered in the 1950s when electrodes were implanted into various brain regions of rats and hooked up to a lever. When a rat pressed the lever, one of the brain regions was electrically stimulated. When the electrodes were placed in the septal area, the rats went wild. One rat pressed the lever nearly 2,000 times per hour—more than once every two seconds, nonstop for an hour. Two decades later a similar study was conducted with a man who had electrodes implanted in three locations and was given a button box with a button to stimulate each of the three regions. Just like the rats before him, he pressed the button for the septal region relentlessly, indicating that it gave him intense pleasure, and he complained at the end of each session when the button box was taken from him.

At the same time that researchers were linking the septal area to reward, other researchers were showing its involvement in fear or, more accurately, its role in reducing fear behavior. One of the best measures of anxiety or fearfulness is called the *startle response*. If someone were to clap her hands loudly behind your head unexpectedly, there would be a cascade of neural, physiological, and behavioral responses that would code this noise as a potential threat and prepare you to respond quickly—a classic fight or flight response. You would probably jump up, turn around, and perhaps notice that your heart was racing. These responses are orchestrated by the amygdala, a phylogenetically ancient structure in the brain often associated with emotional responding. Rats whose septal area has been removed show a much larger startle response and show other evidence of being more reactive to threats. This suggests that when the septal area is intact, it may function to dampen the distress we feel in response to threats.

Last but not least, a separate body of research suggests the septal area is critical for maternal caregiving. Lesion studies in rats, mice, and rabbits suggest that if the septal area is damaged, the animal will be a terrible parent. These lesioned animals no longer make protective nests for their young, they provide their young with less milk, and they experience a much higher rate of infant mortality.

How do we make sense of the various functions of the septal area—reward, fear regulation, and maternal caregiving? Recent work by Tristen Inagaki and Naomi Eisenberger suggests that one way to reconcile the findings is to characterize the septal area as shifting the balance between our approach and avoidance motivations, which promotes proactive parenting. Although humans start planning for their baby's arrival months or even years in advance, most mammals probably do not have the same kind of logical understanding of their relationship to their newborn infants. In the absence of this knowledge in most mammals, screaming babies are a real dilemma. Should we rush to help them or run for the hills? Mammals are wired to fear noisy uncertain things, but the septal area may help to quiet our fears and increase our motivation to help those in need. Instead of selfishly taking cover, we selflessly put ourselves in the line of fire. Thus, the septal region appears to be the key node that converts our affective responses into the motivation to provide help.

It is no accident that this description of the septal area parallels the *nurse neuropeptide* account of oxytocin we saw in the context of social rewards. The septal area is rich in oxytocin receptors, and, for some mammals, this region has the highest density of oxytocin receptors in the brain. Intriguingly, this density is affected by early parenting experiences. Among rodents, pups who receive more parental care grow up to have higher oxytocin receptor density in the septal area, whereas pups who are separated from their mothers grow up to have lower oxytocin receptor density in the septal area.

Empathy is arguably the pinnacle of our social cognitive achievements—the peak of the social brain. It requires us to understand the inner emotional worlds of other people and then act in ways that benefit other people and our relationships with them. It can motivate us to alleviate another's pain or to celebrate someone else's good fortune. All of the neural mechanisms that we have talked about so far need to be coordinated in order to make this amazing achievement possible. Depending on the situation, we need the mirror and/or mentalizing systems to understand someone else's experience. We need the mechanisms that support social pains and pleasures for the affect matching that allows us to feel, not just know, the other's experience. Finally, we need the septal region, central to maternal caregiving, to nudge us to actually get involved in the lives of those around us in positive ways. When all of these mechanisms are in place, we can be our best selves.

Being a Social Alien

In 1992, the year I graduated from Rutgers, I had one of the worst days of my life. As someone who had a long-standing interest in the mind, how it constructs reality, and in science fiction authors like Philip K. Dick who are famous for creating alternate realities, it was more or less inevitable that I would dabble in mind-altering substances as a young adult. If the mind was flexible and reality flexed with it, how could I pass up a firsthand demonstration? There's little point in identifying the particular drug I took that afternoon because the truth was that you never quite knew what you were getting. My roommates from 12 Prosper Street and I headed over to a midday celebration at Cook Campus a few weeks before graduation at Rutgers. We all took the same drug on the way over, something we had all taken before. Everyone had a great time but me. I had the proverbial "bad trip." There was nothing I could do but ride it out.

I have replayed this day in my head so many times, if only to remind myself why drugs are not my friend. In all the times I have revisited this horror film, I never once thought about how I must have seemed to those around me. They couldn't tell what I was experiencing internally. Most had no idea I had ingested anything more powerful than bad keg beer, and those who did know were too busy having fun to care much. I must have come across as very odd, awkward, and antisocial (which, hopefully, is not how I come across the rest of the time). I kept my distance from people, kept my answers very short, and avoided eye contact. I wonder if I didn't seem a bit autistic that day.

Autism is a profound disorder affecting nearly 1 percent of the population. Its dominant symptoms include repetitive behaviors and impairments in social interaction and verbal communication. Asperger's syndrome shares the difficulties in social interaction without the additional language deficits. Clinically speaking, current diagnoses are for *autism spectrum disorders* (ASDs). If empathy is the peak of the social mind, autism is sadly one of its low points. There have been many theories over the years as to why these individuals have so much difficulty with the social world. As we will see, things are sometimes almost the exact opposite of how they seem.

Just two years after the first Sally-Anne test for Theory of Mind was performed, British psychologists Simon Baron-Cohen (yes, Sasha's cousin), Alan Leslie, and Uta Frith proposed that ASD individuals may lack a Theory of Mind. Can you imagine a world in which you didn't see other people's actions in terms of their beliefs, goals, and feelings? Try doing it for a few minutes in your next social encounter. You probably can't, which only shows how much a part of our basic operating system this capacity is. But if you could, it would make you feel a bit like an alien—the bodies in motion around you meaning nothing more than the surface features that you saw. Actions would seem random and unpredictable without your being able to "see" the mind behind them. Could you hold down a job, have friends, or maintain a long-term relationship seeing the world through that lens? Not having a Theory of Mind would seem to explain many of the difficulties individuals with autism have in daily life.

Baron-Cohen and his colleagues tested three groups of people on the Sally-Anne task: children with autism (eleven-year-olds), children with Down's syndrome (ten- to eleven-year-olds), and normally developing children (four- to five-year-olds). Developmental psychologists measure something called *mental age,* which is a way of equating children of different actual ages based on their overall mental ability. The children with autism had the mental age of five-year-olds, hence the much younger age of the normally developing children in the study.

As in previous studies, the normally developing five-year-olds did very well on the Sally-Anne test, with 85 percent of the children getting the right answer. In contrast, only 20 percent of the children with autism passed the test—a radical departure from the results for the normally developing children. Perhaps the performance of the children with autism suffered because of the general cognitive difficulty of the task? If this had been the case, the children with Down's syndrome should have performed poorly as well, but they actually passed at the same rate as the normally developing children. Moreover, the children in all three groups showed a similar ability to recall the facts of what happened during the Sally-Anne task, so it wasn't the case that children with autism weren't able to keep track of the facts. These results point to a relatively specific deficit in the mentalizing ability of children with autism. Subsequent studies have demonstrated other mentalizing deficits in this group, including the inability to make sense of bluffing, irony, sarcasm, and faux pas.

Autistic individuals are also much less likely to describe the Heider and Simmel's Fighting Triangles animations described in Chapter 5 (see Figure 5.1) in terms of mental state characteristics such as beliefs, emotions, and personalities. Ami Klin, a Yale psychologist, reported normally developing and autistic individuals' descriptions of the animations, providing a more concrete sense of their Theory of Mind deficit. First we have a normally developing child's description:

> What happened was that the larger triangle—which was like a bigger kid or a bully—and he had isolated himself from everything else until two new kids come along and the little one was a bit more shy, scared, and the smaller triangle more like stood up for himself and protected the little one. The big triangle got jealous of them, came out, and started to pick on the smaller

triangle. The little triangle got upset and said like "What's up?" "Why are you doing this?"

When you read that description, it is easy to envision social events taking place and how each of the participants in the drama felt. The description is full of mental state language and thus gives the inside story of the minds of the animated shapes. It is natural to describe the scene this way and equally natural to hear someone else describe it this way.

In contrast, here is a description from a child with autism:

The big triangle went into the rectangle. There were a small triangle and a circle. The big triangle went out. The shapes bounce off each other. The small circle went inside the rectangle. The big triangle was in the box with the circle. The small triangle and the circle went around each other a few times. They were kind of oscillating around each other, maybe because of a magnetic field. After that, they go off the screen.

If you read only the second story, you will have no sense of the unfolding of any drama. The child with autism describes shapes moving around with no real meaning, social or otherwise. It's important to note that, strictly speaking, this description is far more accurate than the normally developing child's. The large triangle isn't a bully and isn't jealous. The small circle isn't shy and scared. They are cut-out shapes with no thoughts, feelings, or personalities.

Although the description from the child with autism is more accurate, it is far less *useful*. It doesn't give us the kind of insight we all crave into the psychological drama that unfolded, and it doesn't allow us or the autistic child to predict what might happen next (lawsuits? tire slashings? tearful reunion show?). Not being able to see the world in terms of mental states is a profound disadvantage when everyone else does this naturally. Not only do autistic individuals not see these movements in psychological terms but this also limits their ability to connect and share with others who see these events in a radically different light than those with autism.

The Whole Story?

There is little debate among scientists over whether autism is associated with impaired Theory of Mind abilities. It is. What is debated is, first, whether this impairment is the primary explanation of the real-world symptoms associated with autism and, second, whether autism is caused by Theory of Mind deficits or whether these deficits are the end result of some other developmental process that is not intrinsically tied to Theory of Mind. In other words, is the Theory of Mind deficit a cause or a consequence of autism?

As parsimonious as the Theory of Mind account of autism sounds, few studies have linked Theory of Mind ability to real-world problems in autism outside the lab, and there is increasing evidence that Theory of Mind deficits are not the whole story of autism. There are a couple different reasons for this. Recall that only 20 percent of the children with autism passed the Sally-Anne test. That same fact can be stated in a different way: at least some children can pass this test while still qualifying for the diagnosis of autism. If Theory of Mind were the whole story, then anyone with autism should show a corresponding deficit in Theory of Mind. The Sally-Anne test isn't the only or the hardest test of Theory of Mind, but a subset of autistic individuals with real-world social deficits continue to pass harder Theory of Mind tests, confirming that a person can have autism and yet not have a Theory of Mind deficit.

A second issue is that autistic individuals have other perceptual and cognitive aberrations that bear little relation to the Theory of Mind deficits. Uta Frith gave a group of autistic children and a group of normally developing children an embedded-figures test. One example is given in Figure 7.2. The children were asked to find where in the picture the "hidden" triangle to the left of the baby carriage was present (in the same size, shape, and orientation). I'm guessing you went ahead and did this for yourself and found that it took a little while (hint: it's in the hood that shades the baby). Well, it would take you *less* time if you had autism. Individuals with autism consistently do *better* on this kind of task than other individuals. Improved performance is typically not described as an impairment, but in this case it reflects a kind of cognitive-perceptual imbalance.

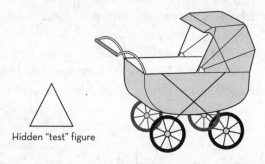

Hidden "test" figure

Figure 7.2 An Example from an Embedded-Figures Test. The exact same triangle (size, shape, and orientation) must be found within the baby stroller. Adapted from Shah, A., & Frith, U. (1983). An islet of ability in autistic children. *Journal of Child Psychology and Psychiatry,* 24(4), 613–620.

The reason why the rest of us are slower at this kind of task is that our minds are designed to focus on the overall Gestalt meaning of what we see, rather than on the details that must be integrated to give rise to that high-level meaning. We see lawns, not blades of grass. Our response to the embedded-figures test depends on our ability to detach from the overall meaning of a picture to search for nonfunctional elements within it. Autism is associated with a deficit in focusing on high-level meaning both in seen objects and in language, and thus if a task requires a focus on detail at the expense of the whole, individuals with autism often excel, outperforming others. There is certainly a parallel between extracting the high-level meaning of seen objects and inferring the goals and motives behind another person's behavior. However, the two deficits do not always go hand in hand in autism, and determining why someone acted a certain way versus why a physical event took place does not rely on the same neural circuits, so these do seem to be distinct processes.

One could still argue that a Theory of Mind deficit is the primary explanation of the social impairments in autism and that Gestalt processing deficits might account for other nonsocial impairments. For this to be true, we would have to assume that training autistic individuals to mentalize better would then lead to corresponding reductions in their social impairments. Multiple studies have shown that training can lead to sizable gains in the mentalizing of individuals with autism, and yet, sadly, those gains produce no real-world improvement in social skills.

Cause or Consequence?

The preceding findings point to clear Theory of Mind deficits in autistic individuals but also suggest that they may not be driving the social impairments seen in autism. This would make more sense if it turned out that Theory of Mind deficits were a secondary consequence of the core deficits in autism, rather than the causal feature intrinsic to autism. To illustrate this difference, imagine a runner who hurts her left knee while training for a race. The odds are quite high that within a week she will also have pain in her right hip. Pain in her left knee will cause her to limp and favor the right leg. This puts undue pressure on those joints and will often cause pain in the right hip opposite to the injured knee. In this case, the knee pain is intrinsic to the original injury, but the hip pain is secondary, acquired because of how the body compensated for the knee pain.

As basic as Theory of Mind is to our adult nature, our acquisition of this ability depends in part on having the right kinds of experiences when we are young. Seeing and hearing other people interact with the world using their mentalizing ability helps to develop our own ability. We know this because children who are born deaf perform just as poorly as children with autism on Theory of Mind tests. The deaf children do not have any mental deficits and are not socially avoidant, but they cannot hear people speaking and thus are exposed to a socially impoverished environment in which conversations about or invoking mental state language are missed. Is it possible that something related has occurred in autism? There is substantial data to suggest that the social deficits in autism precede even the earliest age at which children show evidence of Theory of Mind, so perhaps those earlier changes alter the inputs that autistic children are exposed to.

Autism has historically been diagnosed at the age of three or later. Thanks to the analysis of home movies, we now have a good idea of what these children look like prior to their diagnosis, in the first and second years of life. Using careful systematic coding protocols, scientists can identify differences in children destined to become autistic relative to other children who were not so fated. Children under the age of one already show evidence of poor social interaction and a lack of appropriate social reactions to others. During the second year, these children tend to ignore others, prefer to be alone, and continue to demonstrate poor social skills. If these children are showing a preference for social isolation, it is possible that they, like deaf children, are not getting the social inputs they need in order to develop a mature ability to mentalize on the same developmental schedule as other children. If so, then we would want to look to neural systems that mature much earlier than the mentalizing system.

The Broken Mirror Hypothesis

The mirror system is evolutionarily more primitive than the mentalizing system, given monkeys have mirror neurons but not Theory of Mind. The mirror system is thought to be operational in week-old infants, who show evidence of imitating. If we are looking for a deficit that precedes Theory of Mind deficits and might even lead to them, the mirror system might just fit the bill.

The first hint that mirror neurons might be central to autism was the well-documented problem that autistic children have imitating others. For more than forty years, studies have been conducted in which experimenters have asked children to imitate various behaviors and hand gestures. Children diagnosed with autism consistently perform worse than typically developing children in these studies. Once the mirror system was clearly linked with imitation, the autism-related deficits in imitation led to a handful of neuroimaging studies that resulted in the *broken mirror hypothesis*—that impairments in the mirror system may be the primary cause of autism.

Although early findings have been provocative, it isn't clear how strongly they support the broken mirror hypothesis. For instance, an early study focused on *mu suppression*, a biomarker of mirror system activity measured with an electroencephalogram (EEG). Normally developed individuals produced mu suppression both when observing and performing hand actions. However, individuals with autism produced mu suppression only when they performed hand actions themselves, not when they observed them. Oddly, though, the researchers did not report whether there was a significant difference between the two groups. This might seem like a minor oversight, but without this analysis, the conclusion that mirror system activity differs in the autistic sample is unwarranted.

Two early fMRI studies also suggested atypical mirror systems in individuals with autism, but not in ways that clearly map onto the symptomology of autism. The mirror system has a frontal component (the posterior inferior frontal gyrus and the premotor cortex) and a parietal component (the anterior intraparietal sulcus and the rostral inferior parietal lobule) further back in the brain (see Figure 6.1). One fMRI study found that individuals with autism produced decreased activity in the frontal component of the mirror system but increased activity in the parietal component when imitating facial expressions. The other study found that this group produced decreased activity in the parietal component of the mirror system when imitating hand actions but increased activity in the frontal component. In both of these studies it was clear that the brain was doing something different in individuals with and without autism during imitation, but each was a mix of increases and decreases, and these differences were in the opposite direction in the two studies. Perhaps the bigger issue was that in both studies, the autistic individuals were able to imitate as well as the nonautistic participants, so it was unclear whether the neural effects seen were contributing to any real-world behavioral problems in autism.

Since these early studies, a number of other studies have contradicted them. Multiple studies have shown roughly equal levels of mu suppression in autistic and healthy samples. Similarly, a number of fMRI studies have shown equivalent or enhanced mirror system activity in individuals with autism. So why are these results so messy and inconsistent? If there are imitation deficits in autism, there ought to be mirror system deficits as well, right? Maybe not.

Victoria Southgate and Antonia de Hamilton provided a compelling explanation of why imitation performance cannot be so easily equated with the mirror system in autism. They pointed out that in real life and even in the lab, successful imitating is about more than, well, imitating. To be successful, you need to know what to imitate and when. If an experimenter says "Do this" and then picks up a pen from the table, what exactly are you meant to do? Is the key to pick up the pen? To pick it up with the same hand as the experimenter? To use the same grip or approach it with the same movement through space? The experimenter might code the failure to do any one of these things as poor imitation. Imitating well requires knowing what should be imitated and then being able to implement that precisely. Knowing what to imitate, particularly in lab studies, is about understanding what someone wants you to do. This is a mentalizing task, which depends on Theory of Mind, something we already know is impaired in autism.

Figure 7.3 The Hand Gestures Used During Automatic Imitation. The thumb and fingers making a U shape (left), a wide-open hand (middle), and a closed fist (right).

One way to solve this problem is to eliminate the mentalizing component from the imitation process. Scientists have done this by examining involuntary imitation. Sometimes we imitate others without intending to or even when we know we shouldn't. Imagine the following. You see a hand making a U with the thumb and fingers (see Figure 7.3). When you see the thumb and fingers start to come together into a closed fist, you are meant to do the same with your hand. Next, you see the U-shaped hand again, but this time it opens up as wide as possible, and you are supposed to do that as well. So far, so good. This is a simple imitation task. Here's where it gets interesting. Sometimes you are asked to imitate what you see (that is, to make a fist when you see the target making a fist), and other times you are asked to do the opposite (that is, to open your hand wide when you see the target making a fist).

It is hard to do the opposite of what you see because it's natural to imitate. As a result, nonautistic individuals take longer to perform the incompatible movement (doing the opposite of what they see) than to perform the compatible movement (doing what they see), and this is an indicator of the strength of our automatic imitation tendencies. In the first study to employ the automatic imitation task in individuals with autism, this group not only produced the automatic imitation effect, but the effect was nearly 50 percent stronger in the group with autism compared with the nonautistic participants. Other studies have produced similar effects. Another group using a different automatic imitation paradigm also showed evidence of hyper-imitation associated with autism. Once the mentalizing component is stripped away, rather than failing to imitate, individuals with autism may actually be hyper-imitators.

The broken mirror hypothesis is still relatively new so it's unclear how history will judge it. For the time being, the weight of evidence is against it, with some of the most recent data suggesting that autistic individuals are hyper-imitators who do not know when or exactly what to imitate when explicitly asked to do so. If autistics' imitation difficulty really boils down to having trouble understanding what is to be imitated, this would return us to the well-established finding of deficient Theory of Mind in autism. But we still need to find out why these individuals have this deficit, given that the social impairments exist prior to the development of Theory of Mind abilities, and we know from the study of deaf children that Theory of Mind can develop poorly because of one's experiences alone. Fortunately, we have one more bowl of porridge to try, and hopefully it will be just right.

Intense World Hypothesis

Remember my "bad trip" at Rutgers? There's a second half to that story, the internal half. I described my behavior earlier—what others around me could see. Basically, I kept my distance from everyone and came across as uninterested in my social environment. Obviously I had not become autistic, even for a little while. But there is a lesson to be learned from my experience. I call it the *Head & Shoulders effect*.

You might remember the Head & Shoulders advertising campaign from the 1980s. In these ads there were always two people, one of whom noticed the other's dandruff. The person who saw the other's flakes pointed it out, then said something along the lines of, "Hey, try my Head and Shoulders," and then passed the conveniently on-hand shampoo bottle to the flake-afflicted friend.

The receiver always responded by saying, "But you don't have dandruff!" to which the first individual flashed a winning smile and said, "Exactly." The implication was that because he used Head & Shoulders, no one would ever know he had dandruff. The psychological parallel is that how things look on the outside are often the opposite of what is going on inside the person. We tend to assume that outsides and insides match, but because people react to their circumstances and try to compensate, they often do not.

On that fateful day at Rutgers, people could not have guessed my internal experience from my outward behavior. My behavior and experience were intimately connected, but in a nonobvious way. My outward behavior was antisocial, but this is not because I was uninterested in the social world. Rather, I was *overwhelmed* by the social world. To put a fine point on it, I was overwhelmed by everything, but the social world was the most overwhelming part of all the overwhelming experiences that day.

The drug that I had ingested heightened all my senses. Usually that is pretty cool, but as the guys from *This Is Spinal Tap* might say, my senses were turned up to eleven. It was all just too much for me to process. All the perceptions we have that are usually in the background were suddenly in the foreground and much too intense. An unmoving frozen environment would have been distressing enough, but I was surrounded by people making noise and full of facial reactions, gestures, and other sudden movements. All of it was very intense to me, surprisingly unpredictable, and frankly, terrifying. I was antisocial that day not because of a dislike of those people but rather because of the simple fact of them being people, which made them too much for me to handle that day.

What if children with autism, rather than being insensitive to the social world, are actually too sensitive to the social world? What if the distressing intensity of early social interactions leads these one- and two-year-olds to prefer isolation over social contact? If overly intense experiences promote social isolation, these children might go on to miss out on countless interactions that would train their brains to become social experts over the next decade. Perhaps those with autism are simply cutting class because it is too painful to stay in class. You have probably had the experience of covering your ears when a movie theater demos its sound system before a movie (my wife and son both do this, every time). If your life were always like that, wouldn't you find a quieter place?

I think it is possible that autistic adults are less socioemotionally sensitive, at least in some ways. Even if this is true, the question is whether they were always insensitive to the social world because of their genetic dispositions or whether the lack of social engagement is acquired as a result of the autistic individual's rational responses to a childhood hypersensitivity to the social world. This is the essence of the *intense world hypothesis* of autism. The distress of early life leads these children to turn away from the social world, causing them to miss key social inputs that ordinarily would help the mentalizing system mature.

The intense world hypothesis is relatively new and counterintuitive. Is there any evidence for it? Within the autistic community, people have certainly reported feeling this way. Jay Johnson is a blogger who has autism and writes about this experience in the context of explaining why he does not make eye contact:

> People are loud confusing creatures. . . . And they expect me to add eye contact? I actually don't know how it feels for you, but for me, looking into another person's eyes and having them look back into mine *feels like I am touching a hot stove.* I am being burned. It's like an extra jolt of overwhelming input.

Beyond this sort of anecdotal experience, there is also a fair amount of empirical support, even though few researchers are actually looking to promote this counterintuitive theory. Given the tendency we all have for confirmation bias (searching only for what we hope to see), perhaps this is a good sign for the hypothesis. Early on, it was believed that individuals with autism might be less emotionally sensitive and that part of the reason for this might be diminished amygdala sensitivity. As discussed earlier in the chapter, the amygdala is a small structure that responds to and codes for the emotional intensity of events in our environment. In humans, the amygdala seems particularly responsive to social inputs such as other people's emotional expressions. And while the amygdala does respond to intense positive and negative cues in the environment, there is reason to think it is more central to negative emotional experiences, like fear and anxiety. Even subliminally presented fearful faces that an individual never reports seeing reliably turn on the amygdala.

The best initial evidence for an amygdala-autism link came from studies comparing the neural responses of adults with and without autism to faces expressing emotions like fear or anger. The most consistent finding from these studies was that autistic individuals produced a weaker amygdala response to these threatening social cues. When these findings were combined with the fact that amygdala damage in nonhuman primates can produce some autistic-like characteristics, it seemed to follow that autistic individuals might be somewhat insensitive to the social world because their amygdala wasn't tuning in and directing their attention to the social world.

More recent data focused on children with autism, rather than adults, has suggested a dramatically different relationship between the amygdala and autism. Children with autism actually have *larger* amygdalae than typically developing children. This has been seen in children as young as two to four years old and in children up to age twelve. This is a long time to be walking around with an enlarged mechanism for socioemotional sensitivity, and it is enlarged during one of the most critical socialization periods of our lives.

When we read that a brain region is larger in one group than in another, we assume it must be doing more of whatever it does. Seeing that Einstein had an abnormally large parietal lobe, a portion of the brain critical to spatial skills and mathematical ability, we understand his unfair advantage over the rest of us: he had a bigger better computer back in that part of his brain. However, increased brain volume does not necessarily mean that a brain region is performing better at what it usually does. But for autistic individuals, the enlarged amygdala does actually parallel Einstein's enlarged parietal lobe—more means more.

To the extent that an autistic child has a larger amygdala, that child will also tend to be more anxious—a sign that they may be overwhelmed by the environment. Autistic children also show enhanced threat detection, and their amygdalae do not habituate to faces (that is, calm down with repeated exposures) like the amygdalae of nonautistic children. Critically, increased amygdala volume at age three is predictive of poorer social adjustment later at age six. On top of all of this, the visual pathways that feed potential threat information to the amygdala also show evidence of hyperactivity in autistic individuals. This unusual visual processing in autism may be an advantage when someone is performing perceptual tasks like the embedded-figures task shown in Figure 7.2, but it may also contribute to an overintensity of the inputs reaching the amygdala. Some evidence also suggests that autistic individuals are hypersensitive to sound and touch, in addition to visual inputs.

These results all paint a picture consistent with the intense world hypothesis. This still leaves us scratching our heads about why autistic adults show less amygdala activity in response to seeing emotional faces. Tracking the eye movements of autistic individuals when they are shown pictures of faces gives us a major clue. When you or I see a face, we spend most of our time looking two places—the eyes and the mouth, with our time disproportionately spent on the former. These two spots are especially expressive and convey mountains of information about the emotional state of the other person. When autistics look at a face, their eye movements suggest that they are looking at it very differently. Autistics look almost randomly around the face, often looking at the least informative parts of the face. Nonautistics spend nearly twice as much time looking at the eyes of the target as individuals with autism do.

These differences in social gazing (that is, how we look at faces) suggest that people with autism might show less amygdala activity when looking at faces because they don't look at the emotional parts that typically activate the amygdala. Richie Davidson's research group at the University of Wisconsin tested this idea. They found that when autistic adults looked at the eyes of an emotional target, they showed *greater* amygdala activity than nonautistic individuals, in contrast to the prior studies that had not controlled for the eye movements of the participants. During development, autistic individuals may learn not to attend to sources of emotional information because it is distressing, and thus in adulthood are less responsive because they are deploying this coping mechanism. I don't mean to suggest this is a conscious choice, but many of us engage in self-protective strategies that we have learned through conditioning and have no idea we are using.

Autism is as complex as any known psychological disorder. It has a complex etiology, involving multiple potential causes and developmental pathways. But the intense world hypothesis looks promising. It is counterintuitive because it suggests that what looks from the outside world like insensitivity to the social world is very different from how the autistic individual experiences the world. It suggests that the autistic individual's aversion to the social world is a coping mechanism for dealing with the most intense and unpredictable part of the world (that is, people), which overwhelms them, literally, in each encounter. By missing out on countless social interactions early on, these children lose the opportunity to strengthen their mentalizing abilities during critical periods of brain development. Many of the vicarious inputs that mature our social minds are simply never seen or heard by these children.

Social Cognition

In the three chapters in Part Three we have seen how miraculous our social mind can be, in both the heights it can reach and the alienation it can produce when it does not function in the typical way. Empathy represents the perfect storm of sympathetic sharing of another's feelings, understanding what is likely being experienced and what kind of help or comfort is needed, and having the prosocial motivation to act on behalf of others without necessarily weighing the costs and benefits to oneself. Autism too is a perfect storm, but a tragic one, in which overwhelmed young children may choose to protect their current well-being by forgoing training experiences that develop the mental machinery and facilitate connecting with others more effectively in adulthood.

So far we have seen that social pains and pleasures are real, are present in all mammals, and depend on some of the same neural machinery as physical pains and pleasures. These produce the motivational urge to stay connected with those who can help us survive childhood and thrive the rest of our days. We have also seen the social cognitive machinery that allows us to convert our urge for social connection into thoughtful and enduring relationships among friends, loved ones, and coworkers. Using our mindreading abilities lets us proactively plan for how to get along well with others rather than always being a step behind, reactive and defensive. Our ability to use the mirror neuron system to understand the psychological acts that others engage in, as well as to imitate how those acts are performed, is shared with at least monkeys and great apes. In contrast, our most sophisticated capacity for mentalizing logically is partially shared by great apes and partially exclusive to us.

Now we turn to evolution's third and most surprising bet. The urge to connect and the ability to understand what others think and feel are critical to building an effective social creature. The pièce de résistance is evolution's building us to naturally adapt to the groups we are in and become the kind of people those people want to be around. Here's where evolution got sneaky.

Part Four

Harmonizing

CHAPTER 8

Trojan Horse Selves

n 1641, René Descartes published *Meditations on First Philosophy*, which presented his theory of *mind-body dualism,* later known simply as *Cartesian dualism.* According to Descartes, minds are animated by an immaterial soul distinct from the realm of the physical and all physical processes. There is the mental and there is the physical, and never the twain shall meet. A few decades later, J. J. Becher published *Physica Subterranea,* which similarly focused on an invisible entity. Becher proposed that all flammable materials are ignitable because they contain *phlogiston,* a hypothetical substance without perceivable qualities such as color, odor, taste, or weight. Like Descartes' characterization of minds, fire too is animated by a seemingly immaterial substance in this scheme. Both of these ideas were widely discussed and believed in their day.

Times have changed, and so have the fortunes of these two theories. Whereas mind-body dualism has been one of the most entrenched ideas of the last millennium, informing policy discussions regarding the ethics of cloning, abortion, and the use of animals in laboratory tests, phlogiston is only occasionally mentioned in scientific circles and then only derisively as a cautionary tale of unscientific theorizing. One might naturally assume that the reason Cartesian dualism endures while phlogiston has fallen out of favor is that the former has garnered scientific support while the latter has been refuted by science. But such an assumption would be wrong.

In scientific circles neither theory is reputable, although scientists (including me) still regularly report their findings in dualistic language, referring to the mind as if it is separate from the body. One of the fundamental tenets of the modern science of the mind is that the mind is a thoroughly biological and therefore material entity. Nevertheless, people walk around with an ingrained belief in the simple but implausible form of mind-body dualism that Descartes described. Consider the following dilemma. Would you rather keep your body but no longer have a mind? The body would go on acting just as you do, but the experiencing "you" would be gone. You would no longer have thoughts, feelings, or memories. Or would you prefer to keep your mind without having a body? You would still have experience but no body with which to interact with the world. Any answer to this question is a tacit endorsement of dualism. The truth is that it is easier to think of ourselves as having both a mind and a body that are distinct from one another.

Me in the Mirror

Descartes' belief about our dual nature—mind and body—was a profound error about the way nature works, but it was an accurate assessment of how our brains represent the world. Hundreds of years before the neural data was discovered, Descartes recognized that there is a deep division in these two components of how we see our *self*, body and mind. Does this mean we have two selves? If so, which self do we recognize when we see our reflection?

In 1970, Gordon Gallup made a mirror available to a group of chimpanzees. He was interested in whether chimps were self-aware—if they had a self and knew they had a self. The chimps couldn't talk about themselves, but Gallup thought the way they responded to their own image in the mirror could reveal if they had a sense of self. Chimps are our closest genetic relatives so they are the best candidates to have something resembling our sense of self.

As the chimps spent time with the mirror, they engaged in two kinds of behaviors, suggesting a growing awareness of what the mirror image represented. At first, the chimps acted toward the image as if it were another chimp, someone new in their presence. By the third day, this behavior trailed off dramatically and was replaced by behaviors that were self-focused. For instance, they used the mirror to guide themselves to pick food out of their teeth. After ten days, the key test was performed. While the chimps were asleep, Gallup placed an odorless red dye on each of their foreheads. Later, when the chimps were awake, they were shown the mirror again, and chimps showed clear evidence of self-recognition as they saw the dye in the mirror and touched their own foreheads to investigate. Paralleling work linking social interaction and self-awareness in humans, Gallup also found that chimps who had been raised in isolation never showed evidence of mirror self-recognition.

Not without some controversy, results like these from different species have been used to establish that chimps, dolphins, and elephants are self-aware. Once fMRI became popular decades later, scientists examined self-recognition in the scanner to determine its neural bases. Across more than a dozen studies a clear pattern has emerged. When people see and recognize a picture of themselves, in contrast to pictures of friends, celebrities, or strangers, regions in the right prefrontal and parietal cortex on the lateral surface of the brain are more active (see Figure 8.1). In addition, the parietal region that responds to seeing one's own face also responds to keeping track of one's own body movements.

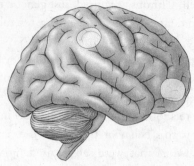

Figure 8.1 Regions in the Right Hemisphere Associated with
Visual Self-Recognition

Neural Dualists

For forty years we have taken mirror self-recognition as a decisive
sign of self-awareness in others, but the truth is more complicated.
In Cartesian terms, this test focuses on the recognition of our body
as our body. In Descartes' meditation, it was the irreducibility of
our minds that led to his famous maxim *"cogito ergo sum"* ("I think
therefore I am"). Long before Descartes, the Oracle at Delphi urged
all to "know thyself," and Socrates exhorted us that "the unexam-
ined life is not worth living." Westerners have taken on this call to
action with increasing intensity over the past millennium. When
we urge people to know themselves, are we talking about the same
kind of knowing that lets us know it's us in the mirror? Mirror self-
recognition is a kind of self-awareness, but is it really representative
of the deep kind of self-knowledge that we seek?

Bill Kelley, Todd Heatherton, and Neil Macrae, prominent social neuroscientists from Dartmouth College, ran a simple but elegant experiment that answered this question conclusively. In an
fMRI study, participants were shown adjectives, such as *polite* and
talkative. For some of the trials, participants had to judge whether
the adjective described George W. Bush, who was the U.S. presi-

dent at the time. On other trials, participants had to judge whether the adjectives described themselves. The critical analysis examined whether there were any regions of the brain that were more active when people judged the applicability of an adjective to themselves as opposed to George Bush. There were only two regions of the brain whose activity followed this pattern.

Just as in the mirror self-recognition studies, there was activity in the prefrontal cortex and parietal cortex. But unlike the mirror self-recognition studies, these activations were present in the medial prefrontal cortex (MPFC) and the *precuneus*—on the midline of the brain where the two hemispheres meet, rather than on the lateral surface of the brain near the skull (see Figure 8.2). In other words, recognizing yourself in the mirror and thinking about yourself conceptually rely on very different neural circuits. Seeing yourself and knowing yourself are two different things. There are at least two major implications of this distinction between self-seeing and self-knowing.

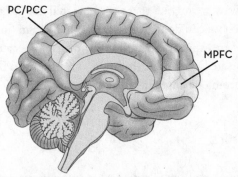

Figure 8.2 Brain Regions Associated with Conceptual Self-Awareness (MPFC = medial prefrontal cortex; PC/PCC = precuneus/posterior cingulate cortex)

First, this distinction clarifies what the mirror self-recognition test tells us about the animals that can pass it. Chimps, dolphins, and elephants all have some sense of their corporeal identity, that the body they see in the mirror is their body. However, the fMRI data suggests that passing this test does not imply that these animals engage in self-reflection the same way that we do, reflecting on whether we possess a particular personality trait or wondering what will become of us in ten years. It does not imply that these animals reflect on the wisdom of their past decisions. And it certainly does not imply that these animals come to have a conceptual sense of self through introspective contemplation.

Second, the neural separation of representing our own bodies and representing our own minds explains why we can't get away from Descartes' mind-body dualism. All signs point toward mind-body dualism being a bad explanation of what we are, and yet most of us operate like card-carrying dualists. We can't help it because it is literally wired into our operating system to see the world in terms of minds and bodies that are separated from one another. We have one system for thinking about our own minds and another one for recognizing our own bodies, and these systems are separated in the brain. It is not that minds and bodies are separate realms in reality, but the ways we register them are separated in our brains, and there isn't much we can do to bridge this neural chasm. Just as colors and numbers are experienced as radically different because they depend on dissociated systems in the brain, our mind and body are forever cleaved from one another.

The Third "I"

The medial prefrontal region that was observed in Bill Kelley's study has appeared again and again in countless studies of self-reflection. These studies demonstrate that our conceptual sense of self is strongly tied to the medial prefrontal cortex (MPFC). In one review that I published, the MPFC was observed in 94 percent of all studies of self-reflection, and it is the only region that is so reliably associated with thinking about "who we are."

Given that we may be the only species able to think about ourselves conceptually, is there something special about the MPFC that allows us to do this? Let's start with a little anatomy to clarify the region we are discussing. German anatomist Korbinian Brodmann examined the cytoarchitectonic structure of cells throughout the human brain near the turn of the twentieth century. He identified about fifty different regions of the cortex, and each has been identified as a *Brodmann area*, a taxonomy that still stands a century later. The medial wall of the prefrontal cortex can be divided into three regions (see Figure 8.3). The ventromedial prefrontal cortex (VMPFC), discussed earlier in the context of reward, is identified as Brodmann area 11. The dorsomedial prefrontal cortex (DMPFC), the central node in the mentalizing system, consists of Brodmann areas 8 and 9. The medial prefrontal cortex (MPFC), identified as Brodmann area 10 (that is, BA10), is sandwiched between the ventromedial and dorsomedial prefrontal cortices. When you point to your "third eye" on your forehead (whether you do so ironically or not), you are probably pointing at this MPFC region, which is responsible for your sense of having an "I." Although there are ongoing debates about the extent to which rodents have something like a prefrontal cortex at all, it is clear that they have no equivalent of BA10; only our closer primate relatives (that is, monkeys and great apes) possess this region at all.

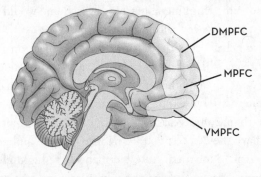

Figure 8.3 Different Regions of the Medial Wall of the Prefrontal Cortex (DMPFC = dorsomedial prefrontal cortex; MPFC = medial prefrontal cortex; VMPFC = ventromedial prefrontal cortex)

Neuroanatomist Katarina Semendeferi examined the size of BA10 across six of the primate species, including humans, that have it. The size of this region was below 3,000 cubic millimeters in chimps, bonobos, gorillas, orangutans, and gibbons; it was above 14,000 cubic millimeters in humans. As we saw earlier, raw size comparisons aren't terribly meaningful because the human brain is so much larger in general. What is more informative, then, is the percentage of the total brain devoted to BA10. In nonhuman primates, BA10 takes up between 0.2 and 0.7 percent of the total brain volume. In humans it takes up 1.2 percent of the total brain volume. Put another way, BA10 takes up twice as much space in the human brain as it does in the chimpanzee's brain. BA10 is one of the only regions of the brain known to be disproportionately larger in humans than in other primates. Semendeferi also discovered that BA10 is less densely populated with neurons than other cortical regions. Reduced crowding gives each BA10 neuron space to connect to a greater number of other neurons.

The MPFC is clearly pretty special, distinguishing us from other primates. Given that humans are the only species that we know for sure have a conceptual sense of self, it makes sense that this capacity would be linked to a brain region that is distinctive in humans. So what does this region do for us? We in the West spend a lot of time thinking about ourselves. Some would go so far as to say we are obsessed with ourselves.

We believe that the self that we introspect on is composed of our private stock of personal beliefs, goals, and values. It holds our hopes and dreams that no one has access to but us. Chinese philosopher Lao Tzu captured the idea of the self as a source of truth more than 2,000 years ago, writing, "At the center of your being you have the answer; you know who you are and you know what you want." Nobel Laureate Herman Hesse highlights the distinctiveness of each self from the others, arguing that each "represents the unique, the very special and always significant and remarkable point at which the world's phenomena intersect, only once in this way, and never again." If the self represents our unique nature, then the MPFC appears to be the royal road to knowing our own hidden truths and the best route to securing personal happiness. But as we have seen, things are not always as they first appear.

Trojan Man

The myths of Virgil and Homer tell us that Helen of Troy was taken from Greece by a Trojan named Paris in the thirteenth century BC. Agamemnon, Helen's brother-in-law and Greek royalty, led a siege against Troy that lasted a decade. The Trojans withstood the frontal assaults, never allowing the Greeks to breech the city limits. The Greeks finally turned the tide with the well-known stratagem of the Trojan horse. The Greeks staged a hasty retreat, leaving behind a giant wooden horse, which the Trojans wheeled into the city as a trophy of their victory. Unbeknownst to the Trojans, there were Greek soldiers hiding inside the horse, silently waiting for nightfall. After dark, these warriors exited the horse, took the Trojans by surprise, and opened the gates to the Greek army, resulting in a speedy end to a long war.

Why this digression into Greek history? The Trojan horse was not at all what it seemed. Instead of being the spoils of war, it was a cleverly disguised deception that allowed the Greeks entry into Troy and led to its being overtaken by the Greeks. In the same vein, I would argue that we can describe our sense of self as a *Trojan horse self*. In the West, we like to think of the self as that which makes us special, providing us with a unique destiny to reach our personal goals and achieve self-fulfillment. We imagine the self—our sense of who we are—to be a hermetically sealed treasure chest, an impenetrable fortress, that only we have access to. If this were really the whole story, such a discussion of the self wouldn't have a place in a book on the social brain. But as it turns out, the self may be evolution's sneakiest ploy to ensure the success of group living. I believe the self is, at least in part, a cleverly disguised deception that allows the social world in and allows us to be "overtaken" by the social world without our even noticing.

Nineteenth-century philosopher Friedrich Nietzsche gave the most cynical view of this Trojan horse self, writing:

> Whatever they may think and say about their "egoism," the great majority nonetheless do nothing for their ego their whole life long: what they do is done for the phantom of their ego which has formed itself in the heads of those around them and has been communicated to them.

Nietzsche believed that our sense of self was not something inherently internal to us, a true core to our being, that we gained greater access to over the course of our lives. Instead, he argued that our sense of self is typically something constructed, primarily by the people in our lives, and that the self is actually a secret agent working for them more than for us. If one believes that the purpose of the self is to help each of us of maximize personal rewards and achievement by better knowing who we are, then it would be tragic to discover that our self actually does something very different from what we think it does.

Our responses to cultural trends give us some insight into how this process works. When I see a new look in clothes, my first reaction is often "that looks ridiculous," yet a few months later I find that the trend looks and feels "right" to me. For a dramatic example of this, consider baby colors. Go to any store that sells baby supplies, and you will see a wide array of clothes and equipment in blue or pink, for boys and girls, respectively. At one level, I don't like that boys and girls are already separated this way from birth. At another level, I get it. Blue for boys and pink for girls just feels right. It may not be PC, but it's right—I can feel it in my gut. Just imagine if some store tried to switch things up and sell pink for boys and blue for girls. That would never catch on, right? Actually, it already did.

A hundred years ago, the color scheme for babies was the opposite of what it is now. Consider this comment from a trade journal published in 1918:

> The generally accepted rule is pink for the boys, and blue for the girls. The reason is that pink, being a more decided and stronger color, is more suitable for the boy, while blue, which is more delicate and dainty, is prettier for the girl.

Somehow between 1918 and now, our visceral reactions have done a full reversal. Imagine that in the 1920s some trendsetters decided to assign blue to boys and pink to girls. I'm sure they must have been laughed at initially and yet somehow the change caught on. Over time, everyone's associations slowly changed until blue for boys went from seeming so wrong to seeming so right. Did each person reach this conclusion privately, or was there some process at work to make sure our way of seeing things stayed in line with what we perceived to be the beliefs of those around us? Just like most of our beliefs, this visceral response to baby colors is something we pick up from the outside without even noticing it. I don't mean to imply this happens 100 percent of the time for 100 percent of the people. It doesn't. But it is odd how frequently and easily we shift our attitudes along with the masses.

I have argued that evolution is moving us ever closer to interdependent social living, where we maximize what we can do together in groups. If that is the case, having our beliefs and values injected into us from the outside in a "clandestine operation" would yield greater *harmonizing* among people in groups and lead to an improved balance of social pains and pleasures. Each of us has a variety of impulses—desires that if acted on at the wrong time, in the wrong place, and with the wrong people, could bring civil society to its knees.

I would argue that the self exists primarily as a conduit to let the social groups we are immersed in (that is, our family, our school, our country) supplement our natural impulses with socially derived impulses. The social world imparts a collection of beliefs about ourselves, about morality, and about what constitutes a worthwhile life. Because of how the self functions, we often cling to these beliefs as though they are unique ideas we came up with for ourselves—the products of our private inner voice. It is not enough for us to recognize what the group believes and values. We have to adopt the beliefs and values as our own if they are to guide our behavior. In other words, just like the Trojan horse, much of what makes up our sense of self was snuck in from the outside, under the cover of darkness. We might believe the self exists to help strengthen our resolve in the face of outside forces, but this theory of "who we are" overlooks the ways our brain uses those outside forces to construct and update the self.

In Your Eyes

Imagine you are sitting in a room with twenty people, and each of you is handed a card from a standard deck of playing cards. You are not allowed to look at the card; instead, you hold it on your forehead for all others to see. Everyone in the group is then told to try to "pair up with the person with the highest card that will pair with you." In the beginning, you can see everyone else's "value" but have no idea of your own. Within moments you will have a pretty good idea though. The woman with the ace of hearts will have a crowd of admirers, hoping to be chosen by her, while the man with the two of spades will quickly realize why no one is returning his gaze.

George Herbert Mead and Charles Cooley, influential psychologists in the early 1900s, suggested that learning about ourselves in the real world is not so different from this little card game. In many cases, it is hard to look inside and really know who we are, and thus we tend to look to others, both intentionally and unintentionally, to find out. Mead and Cooley discussed a process that later became known as *reflected appraisal generation*. At its simplest, a reflected appraisal is what I think you think of me. We are bombarded with feedback from others about ourselves, sometimes in words but more often in the form of their nonverbal behavior and tone of voice. Mead and Cooley argued that we use this information to find out who we are. Rather than looking inward, we often look to others to learn about ourselves. If psychologists want to study the self while it is still under construction, when a person is actively generating these reflected appraisals to flesh out their conceptual sense of self, we ought to be looking at the adolescents who devote a lot of time and energy to this. When Jennifer Pfeifer was a graduate student in my lab, she convinced me to do just that.

We asked young adolescents (that is, thirteen-year-olds) and adults to report on both their *direct appraisals* of themselves (for example, "I think I am very smart") and their *reflected appraisals* (for example, "My friends think I am very smart"). There are a few things we should naturally expect to see in this study. First, direct appraisals should activate the MPFC, given that Bill Kelley and others have shown this link in dozens of studies. Indeed, this was the case for both adolescents and adults (though adolescents showed greater activity than adults, consistent with adolescence being an intense period of self-focus). Second, reflected appraisals should activate the mentalizing system as these involve figuring out what somebody else believes. This too was observed in both adolescents and adults.

The results got more exciting as we moved into uncharted territory. Before this study, no one had ever examined how a thirteen-year-old's brain makes sense of itself. Our adolescents produced strong activity throughout the mentalizing system while making direct appraisals of themselves. The adults did not. Recall that the mentalizing system is typically associated with thinking about the mental states of others. These results suggested that even when we asked adolescents what they thought of themselves, they might have spontaneously brought to mind reflected appraisals, what they believed others believed about them. Rather than answering by directing their thoughts inward, adolescents may have unwittingly been focused on the minds of others when answering about themselves.

The other novel result from this study reinforces this idea of self-knowledge constructed from outside sources. The adolescents in the study activated the MPFC both when making direct appraisals and when making reflected appraisals. This is important because on the surface, direct and reflected appraisals are very different psychological processes. A reflected appraisal is my assessment of what you believe—a standard mentalizing task that might not be related to my internal experience of myself. In contrast, a direct appraisal feels like it taps into a personal place of self-truth that only I have access to. Yet here we saw the MPFC involved in both. Perhaps in coordination with the mentalizing system, the MPFC is taking our assessments of what others believe about us as a proxy for what we should believe about ourselves. If this is true, then the medial prefrontal cortex is not the royal road to personal truth but rather a reflection of various sources by which we learn about ourselves— some personal and introspective and some generated from what we believe those around us think about us. This suggests that the MPFC may be involved in a social construction of the self. But is the MPFC actually involved in others' influencing us and changing our beliefs?

Changing My Mind

In my twenties, I was a rabid fan of the Blue Man Group stage show (I am still a big fan, but the rabies have passed). I've seen the show at least a dozen times, in New York City, Boston, Las Vegas, and Hollywood, and I have taken hundreds of people to see the show with me. I even auditioned to become a Blue Man myself when things weren't going so well in graduate school—my version of trying to run away and join the circus. If you haven't seen the show—go see it. I'll wait. The Blue Men are essentially aliens who have landed on our planet, trying to make sense of who we are, and they are trying to connect with the audience through various means. But the Blue Men are mute, and have their own inimitable way of doing things.

One of my favorite parts of the show involves a woman from the audience being brought up on stage. The selected woman is always young and pretty, and more often than not, she is wearing a white sweater. Once on stage, she sits between the Blue Men at a long table, participating in a scene that involves the Blue Men vacuuming pictures of furniture off a painting, eating Twinkies, and then "spitting up" all the food from a valve in a chest plate they each wear. Throughout, each Blue Man works hard to curry favor with the girl, flirting as only Blue Men can, each trying to "one up" the others.

It's a very funny scene with lots of expressive nonverbal behaviors (remember, Blue Men are mute), and the orchestration of the interactions is exquisite. People assume that the girl is planted in the audience and works for the show because there is no way, without verbal instruction, she could hit all her marks in the scene so precisely. After all, the Blue Men are not above sleight of hand. Once, when I was pulled up on stage, a headphone was surreptitiously placed in my ear to give me instructions. But years later, I got a chance to meet the Blue Men—Chris Wink, Matt Goldman, and Phil Stanton—and they assured me that in the Twinkie skit, the woman is always a regular audience member.

The skit worked because we humans are built to be influenced by those around us, to follow their lead. In other words, we are far more suggestible than we know. Each Blue Man behavior elicited an appropriate preordained response from the unwitting female accomplice. In the West, we call this conforming, something looked down upon. But in the East, the same behavior is called *harmonizing*, something essential for successful group living.

Suggestibility and the process of being persuaded have been

studied in a few different ways with fMRI. If the MPFC not only represents our sense of self but also opens the gates to the Trojan horse self, allowing those around us to influence us, then the MPFC ought to be involved in suggestibility and persuasion. Despite our intuitive sense that knowing ourselves is what keeps us from being unduly influenced by the social world, the MPFC is actually central both to self-knowledge and to being influenced by others.

If you have never been hypnotized yourself, you have probably seen someone else being hypnotized. Hypnosis is real, though most people are not deeply hypnotizable. For the few that are profoundly hypnotizable, color images can literally be made to appear colorless, surgeries can be performed with no anesthetic, and lifelong smoking habits can be erased in an hour. Amir Raz conducted an fMRI study examining the neural difference between people who were highly suggestible when hypnotized and those who were less so. He had them perform a *Stroop task* in which they were shown color names (for example, R-E-D) that were printed either in the same color ink as the named color or in a different color ink. On all trials, participants were asked to select the ink color that the word was written in. It is well known that people are faster to identify a word written in blue ink if the word spells out B-L-U-E than if it spells out R-E-D. Raz found that if highly suggestible individuals were given the hypnotic suggestion to see the words as nonsense words, instead of color words, their reaction times on the mismatch trials would speed up. In other words, Raz was testing the hypothesis that if a person no longer saw the word written in blue ink as spelling R-E-D, then it shouldn't produce the conflict that usually slows people down in this task. On these incongruent trials, the highly suggestible individuals were much faster than the less suggestible individuals. When Raz looked in the brain to see which brain regions responded differently in the two groups of individuals, the MPFC was one of the central regions observed.

Although people are rarely influenced through hypnosis in their daily lives, attempts to influence through other means are everywhere. We are bombarded with persuasive messages everywhere; advertisements take aim at us through every form of media. Emily Falk and I have conducted a series of studies to examine how other people's opinions cross the blood-brain barrier and influence us every day to behave more in line with those opinions. We were particularly interested in whether the brain contained information about this persuasion process that people could not consciously report on. If so, this would suggest the Trojan horse self was particularly stealthy, influencing us without our awareness.

In our first study, we convinced undergraduates at UCLA to use sunscreen more often. Given that Los Angeles is technically a desert, daily sunscreen use is a good idea around here. We brought students into our scanning facility and asked them a series of questions about their attitudes and recent behavior. Mixed in among the questions were items about how much they had used sunscreen in the last week, how much they intended to use it in the next week, and to what extent they believed that people should use sunscreen regularly. Then the person would get in the scanner and see persuasive messages about sunscreen use from places like the American Association of Dermatology. After leaving the scanner, the individual was once again asked a series of questions, including two assessing their intentions to use sunscreen in the next week and general beliefs about regular sunscreen use. A week later, we contacted each participant out of the blue to find out how many days in the previous week they had actually used sunscreen.

After scanning, some folks told us they had "found religion" and would start using sunscreen every day. Other people said thanks, but no thanks, and they planned to continue on in their sunscreen-free ways. The relationship between what people said they would do and what they actually did was negligible. One would think that if people changed their stated intentions after seeing the persuasive messages, this would be a good indicator that they would really change their behavior. But we need only think of our failed New Year's resolutions to know intentions do not always become reality.

In our study, some people increased their sunscreen use, and other folks didn't, but their actual behavior bore little relation to what folks told us they intended to do. Just looking at their self-reports and their behavior, it all seemed a bit random.

In contrast, participants' brain activity in the MPFC while seeing the persuasive messages predicted their sunscreen use over the next week quite nicely. The more participants' MPFC was activated in response to the persuasive messages, the more those individuals increased their sunscreen usage later on, regardless of what they told us they planned to do. The activity in this brain region did a much better job predicting their behavior over the next week than anything the participants consciously told us. Relating this back to the idea of a Trojan horse self, this study shows people changing their mental representations of the value of using sunscreen in a way that drives behavior but, at the same time, in a way that they are unaware of. People didn't realize the actual change that was taking place within them. And the site of this change in the brain is the MPFC. Once again this suggests that this thing we call our "self" is far less private and hermetically sealed off from the rest of the world than we believe. As it turns out, the way our MPFC responds to an advertisement not only predicts how we will change but also how entire populations will change.

Neural Focus Groups

John Wanamaker, a nineteenth-century pioneer of retail sales, once quipped, "I know I'm wasting half of my advertising budget. . . . I just don't know which half." Ever since, people have been trying to predict which advertising campaigns will succeed or fail before committing their advertising dollars. In truth, we aren't very good at figuring this out because the typical method involves asking a "focus group" what they think. Does this ad make you want to buy the product? Do you think it will make other people want to buy it too? Which of these two spokesmen made you want the product more? Focus groups don't work all that well because people don't actually have introspective access to the answers to these questions. Using a focus group might be better than throwing darts at a dart-board in a dark room—but not much.

Based on our sunscreen study, Emily Falk and I suspected that it might be possible to create a *neural focus group* from which we gathered neural responses elicited in response to ads in order to predict how successful the ads would be when they were aired on television. Our first step in doing this was to replicate the sunscreen study, but this time we used antismoking ads shown to people who were about to attempt to quit smoking. We measured the amount of carbon monoxide in their lungs the day we scanned them (that is, before quitting) and a month later as a biological measure of how much they were smoking at each point in time. Our sunscreen results replicated beautifully: while participants watched the anti-smoking ads, activity in the same region of the MPFC predicted successful smoking reductions much better than the participants' self-reported beliefs and intentions.

The next thing we did was separate the ads based on the adver-tising campaign that they came from. The ads came from three dif-ferent campaigns that had aired in different states at different times during the year. I'll call them campaigns A, B, and C. Simulating a focus group, we asked each of our smokers which ads would be most effective in helping smokers quit. They told us that campaign B was the best, followed by A, with C coming in last. But when we looked at the activity in the MPFC in response to each ad campaign, we saw a very different pattern. The participants' MPFC responded most strongly to campaign C and least strongly to campaign A. In other words our subjects told us that the ads from campaign C were the worst, but their brains told us these same ads might in fact be the most effective.

How could we tell which was right—people's words, their MPFC responses, or perhaps neither? Luckily, each of the ads ended with a specific request of viewers: "Call 1-800-QUIT-NOW." This is the National Cancer Institute's antismoking hotline, and through our public health partners on this project, we were able to find out how many people called this number after seeing one of the ad campaigns. As it turned out, people's MPFCs were prescient. Each of the ad campaigns was successful, but they differed in how much. Campaign B, the one people said would do best, increased the number of calls tenfold. Campaign A, the one people said would do next best, doubled the number of calls that came in. But Campaign C, the one that people said would do worst and that MPFC "said" would do best, actually increased call volume more than thirty times over. Campaign C was more than three times better than the next best campaign.

In addition to reaffirming the common finding that people aren't very good at predicting their own or other people's behavior, we found something of an antidote to the typical misinformation obtained when people make these predictions. People may not be able to consciously tell you what they or others will do in the future, but their MPFC can sometimes provide more accurate predictions. There are times when the brain contains hidden wisdom that if monitored could help us in various ways, whether in marketing, in lie detection, or even in predicting daily stock market fluctuations. People might not "know" these things, but it's possible there is diagnostic information waiting to be uncovered in the folds of our brains, the most sophisticated computer in the known universe.

The second thing this study did was put a nail in the coffin of the idea that our self is what makes us distinct from others. From both the hypnosis and the sunscreen study, we have seen that the region of the brain that is so strongly linked to our conceptual sense of self is also the superhighway by which others influence our beliefs and behaviors. In the neural focus group study, rather than representing what makes us unique and different from others, the MPFC is actually serving as a proxy for how countless others will respond—hardly a marker of our uniqueness.

Not So Self-ish

If the MPFC is in fact a conduit for us to assimilate the values and beliefs of those around us, then the self may truly be a mechanism of, and for, the social world. The existence of the MPFC, more than any other mechanism in the social brain, ensures that a common set of values is largely shared by those in a long-standing community. The MPFC-mediated self may be the mechanism by which cultural norms and values are likely to flourish—lodging notions in our heads that we are committed to before we realize it, so that they are part of the common background of our identities and beliefs.

Although the adolescent years might be a time when we're particularly self-absorbed, most of us eventually embrace an identity that centers on our relationships to friends and loved ones, as well as on the various groups to which we are connected (for example, religious, political, or athletic groups). Once we stop trying to define ourselves exclusively in terms of our uniqueness and accept a more balanced social identity, we often feel that we are finally who we were meant to be. As philosopher Alain de Botton wrote, "Living for others [is] such a relief from the impossible task of trying to satisfy oneself." Albert Einstein conveyed the same sentiments decades earlier, "Only a life lived for others is a life worth while." In an interview, the comedian Louis C.K. similarly described how his identity changed after having kids:

> I don't really remember what it was like before. Whatever I had going on, it was bullshit. It wasn't important. It's kind of a nice thing about being a dad. My identity is really about them now, and what I can do for them, so it sort of takes the pressure off of your own life.

I have certainly experienced this in my own life. Having a wife and a son has given me great focus and clarity on what matters to me. Nothing I came up with on my own before having them in my life ever came close to giving me the same sort of solid stable identity. The modern world has created an extended period of adolescence and self-discovery, and thus, we think of this search for our unique identity as the most natural thing in the world. But I'm not so sure the self evolved primarily so that the Marilyn Mansons and Lady Gagas of the world could make a living out of being as different as possible from everyone else. Prior to the modern era, humans spent a few years being cared for as children and then moved into the workforce, often with responsibility for others, by the teenage years. Most had no time for soul searching—life was about being cared for or taking care of others from beginning to end.

Each of us is a blend of the distinctive and the common, the unique and the shared. But we often think of ourselves in a pitched battle between being true to our self, which is all about standing apart from the crowd, and our need to fit in, which causes us to conform against our wills. In a commencement speech, Steve Jobs warned the new graduates not to let the "noise of others' opinions drown out your own inner voice" but rather to "have the courage to follow your heart and intuition." The data we have seen focusing on the MPFC suggests that this isn't the right story to be telling. Our sense of self, our "heart and intuition," is actually part of what ensures that most of us will conform to group norms, promoting social harmony. Our self works for the group to ensure that we will fit in. This may not have been true for Steve Jobs, but it is true for the vast majority of people. We have selfish impulses and we have socially created beliefs and values that are also internalized as part of the self. There may be a battle between them, but by the time the battle is happening, it usually isn't us against them. It's us against us—two parts of our own identity duking it out. Fortunately, evolution had one last trick up its sleeve to help the socially internalized impulses win their battles against our more self-interested impulses.

CHAPTER 9

Panoptic Self-Control

I am a psychologist, whether I'm on the clock or not. Psychology is the filter through which I see life, read books, and watch reality television. It may come as little surprise then that I occasionally examine my son's social brain development a bit more explicitly than most other people do. I haven't put him in an fMRI scanner or attached EEG electrodes to his scalp (yet!), but I do pay attention to various milestones commonly associated with the maturing of the social brain. Babies have been shown to imitate their parents almost from the moment of birth, but our son, Ian, didn't imitate for the better part of his first year. On the other hand, babies typically pass the mirror self-recognition test at around two years, whereas Ian was obsessed with his own reflection by the six-month mark. When Ian was two and a half, he passed a Batman–Iron Man variant of the Sally-Anne false belief task, but we failed to replicate that Theory of Mind result in several additional tests. My favorite study of Ian's social brain development was definitely the *Popsicle test*.

We live an hour away from Disneyland in Southern California, so Ian was a veteran at a very young age. When he was two, we took him to Disney, and despite being the first ones into the park at 8 a.m., we could barely drag him out of the park at 11 p.m. Not only was that day clearly the best of his 800 days of life up to that point, but I would be willing to wager that it had the greatest volume of sheer joy he will experience in a single day for the rest of his life.

When Disneyland came up with their tagline that it's "the happiest place on earth," they clearly had Ian in mind.

A month before Ian turned three, we asked him whether he would rather have a birthday party or go to Disneyland for two days. It took him all of two nanoseconds to answer. The night before his birthday, he was excited to go, and it was obvious that there was nothing he wanted more than to get to Disneyland the next day . . . or so I thought. Thus began the Popsicle test. He had just finished dinner when he asked for a Popsicle for dessert. Naomi got the Popsicle out of the freezer, unwrapped it, and was about to hand it to him before I stopped her.

"Ian, where are we going tomorrow?" I asked

"Disneyland!!!!" he replied with intense excitement, arms waving in the air.

Ian stared intently at the Popsicle while I asked the next question: "Ian, if you could just have one of these two things, which would you rather have? Would you rather have this Popsicle right now or go to Disneyland tomorrow? If you could only have one of them, which would you choose?" We have video of this episode, and the first thing you can see after I ask this question is a moment of existential dread on Ian's face. He gets it. He wants both of these things intensely but can have only one. The moment evaporates, transforming quickly into his cheerful reply, "The Popsicle!!!"

Despite Disneyland's being Ian's favorite place on earth, he was willing to give up that entire day for the Popsicle that was compelling more for being right in front of him than for the brief and modest pleasure it would actually give him. He could not resist a pleasure in the here and now, no matter how tepid it was, compared to its alternative the next day. Yes, we still took him to Disneyland (we aren't cruel), and, yes, it was much better than the Popsicle.

You may recognize this as a modern-day variant of Walter Mischel's famous *marshmallow test*, pitting a smaller immediate reward against a larger later reward. In the 1970s, Mischel tested preschoolers between the ages of three and five on their ability to wait for a more desirable reward when a less desirable replacement could be had with no delay. The best-known variant involved marshmallows and a bell. The children sat at a desk and were told that the experimenter was leaving the room and that if they could wait until he returned (fifteen minutes later), they could have two marshmallows. However, the children also had the option at any moment of ringing the bell to alert the experimenter to return early, at which point they could have one marshmallow (but not two).

Let the willpower games begin. None of these children were on a diet, so to them, the more sugar the better: they all wanted two rather than one marshmallow. Despite their intention to last the full waiting period, less than a third did so; the temptation was too great. On average, the children lasted about five minutes before giving in. Over the years, Mischel found ways to help the children last longer in their quest for the most marshmallows. Replacing the actual marshmallows with pictures of marshmallows dramatically increased how long the children could wait. In essence, children were better able to resist the idea of marshmallows than real marshmallows (even though the reward was real marshmallows in both cases). Symbolic replacements are less tempting than the real thing. Mischel also investigated how the children could mentally approach the task in different ways to improve their odds of success.

Even when the marshmallows were left on the table, the children were able to demonstrate impressive waiting abilities if they were given tricks for how to think about the marshmallows. Mentally focusing on aspects of the marshmallows that had nothing to do with eating them (for example, "marshmallows are the same colors as clouds") increased waiting times considerably. Shockingly, if the children just imagined that the marshmallows in front of them were in a picture rather than actually there, they were able to wait three times as long as when they were looking at a picture of marshmallows but pretended they were real. Under the right circumstances, the power of mind is remarkable.

All of Life's Good Stuff

In the United States, there is no more pivotal moment in determining ambitious adolescents' future than finding out which college they did or did not get into. Getting into Georgetown rather than Greendale will open a variety of doors, confer higher prestige, and typically lead to a higher paying job. This, in turn, improves one's dating pool, the house one can buy, and the vacations within reach. For most students, high school grade point average (GPA) and Scholastic Aptitude Test (SAT) scores disproportionately influence the odds of getting into any particular school.

Ability to delay gratification makes remarkable contributions to both GPA and SAT scores. Mischel retested his preschoolers once they had taken the SAT years later. Those who had lasted longer at age four scored better on the SATs. In fact, preschoolers who could wait until the experimenter returned of his own accord scored more than 200 points better on the SAT than those who gave up after thirty seconds. More recently, Angela Duckworth found that GPA was better predicted by a person's ability to delay gratification than by their IQ.

Since the discoveries linking self-control and academic outcomes, various other findings also point to self-control as the key to the good life. People with higher levels of self-control have higher incomes, higher credit scores, better health, and better social skills from childhood to adulthood, and they report being happier with life.

Self-control is clearly one of the greatest assets a person can have, but you might be wondering what it has to do with the social brain. We will get there, but before we do, let's be clear about what we mean when we talk about self-control. Self-control typically involves some impulse, urge, or reaction that we want to stop or prevent. Our impulses and emotional reactions are essential in guiding us toward desirable outcomes and away from danger, but they also seem to have a mind of their own and often need to be restrained. Whether it's avoiding that extra slice of pizza at 2 a.m., not telling your boss what you really think of him, or overcoming the urge to drive on the right side of the road when you are visiting London, your habitual responses need to be put in their place from time to time.

When you feel yourself exerting effort targeted at overcoming one of these undesired responses, that's self-control. Why does self-control relate to GPA? Probably because kids who can keep the urge to play videogames at bay long enough to do their homework will do better in school. Why does self-control improve SAT scores? In part, because self-control helps a person to persist in the face of the colossal boredom that is the SAT test and all the preparation that goes into getting ready for it. And rather than going with the first answer that comes to mind and moving on to the next problem, which is driven by the impulse to be done with the test, self-control allows the students to stay focused on each problem until they are sure they have the best possible answer.

One of the defining characteristics of self-control is that it seems to be a limited resource. Essentially, we can engage in only one kind of self-control at a time. Try to actively control two things at once (for example, resisting pizza and memorizing a poem for class), and one or both of these efforts will most assuredly suffer. More surprisingly, engaging in two forms of self-control in sequence can be problematic as well. Trying very hard not to laugh during a funny scene right now will actually make it harder for you to stay focused while taking an analogies test five minutes from now.

In order to explain the discovery that self-control exertion now can undermine self-control later, social psychologists Roy Baumeister, Todd Heatherton, and Katherine Vohs have theorized that self-control is like a muscle. They argue that this effect occurs because the self-control muscle can get fatigued and needs time to recover. Similarly, a muscle can do only one thing at a time. Just like a muscle, self-control is powerful but limited. The muscle perspective has also been extended to suggest that self-control can be strengthened by exercising it. Lifting weights depletes our muscles in the short run but makes them stronger in the long run. The same may be true of our "self-control muscle."

The Brain's Braking System

Part of what makes the sequential self-control findings surprising is that the kinds of self-control we engage in are so different from one type to the next that it is hard to believe that each of these really depends on the same processes in the brain. What does holding back laughter at a comedian's jokes have to do with the focusing you do when you take an analogies test? Why should driving on the left side of the road to get to a business meeting in London affect your ability to keep your cool if the meeting goes poorly?

Although various mechanisms are at work in the different kinds of self-control we exert, one mechanism seems to be at work in nearly every instance. The *ventrolateral prefrontal cortex* (VLPFC) of the brain (see Figure 9.1), especially in the right hemisphere (rVLPFC), activates reliably in numerous types of self-control exertions, irrespective of how different our experiences of self-control feel from one to another. It is the only region in the prefrontal cortex that is larger in the right hemisphere than in the left, but this asymmetry doesn't emerge until late adolescence, when self-control skills significantly improve. For these reasons, it is appropriate to characterize the rVLPFC as the central hub of the *brain's braking system*. Let's take a little tour of some of the diverse ways we engage in self-control, and let's examine the VLPFC's involvement.

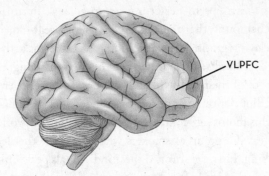

Figure 9.1 Ventrolateral Prefrontal Cortex (VLPFC)
Involved in Self-Control

Motor self-control. The psychologist's favorite motor self-control task is the *go/no-go task* (the *stop-signal task,* discussed in Chapter 3, is a variant of this task). In it, participants are shown an endless series of letters, about one per second, and they are asked to press a button as quickly as possible when each letter appears. They do this for every letter except for one that is predesignated as the *no-go letter.* Whenever this letter appears, the participants do nothing at all (that is, "no-go"-ing). The task is difficult because the no-go letter appears only 15 to 20 percent of the time, so participants ease into the habit of pressing the button once every second or so, and this prepotent motor response has to be overcome when the no-go letter appears. Although subjects are instructed to do nothing in the no-go trials, restraining themselves from pushing the button feels like more work than other trials.

Countless studies have observed increased activity in the rVLPFC (also called the *right inferior frontal gyrus*) when individuals successfully avoid pressing the button in no-go trials. One study examined patients with different kinds of brain damage and found that only brain damage in the rVLPFC was associated with deficits on the no-go task. Decades after Mischel performed the initial marshmallow tests, a subset of his preschoolers were brought back in as grownups to perform a go/no-go task in an MRI scanner. Mischel found that the adults who as four-year-olds had been best at delaying gratification also produced the most activity in the rVLPFC as adults, suggesting that this response might have been at the root of their real-world self-control successes over the years.

Elliot Berkman and I tested the idea that rVLPFC activity during this motor self-control task is a proxy for real-world self-control. We tested a group of smokers who had made the decision to quit smoking. First, we scanned these individuals performing the go/no-go task the day before their "quit day"—the day they had chosen to start trying to quit. This battle for self-control over an intense undesired habit consists of an endless series of skirmishes, in which our urges and our better angels clash several times each day. We wanted to see whether the rVLPFC played a role in tipping the advantage to self-control.

To get at the moment-to-moment self-control conflicts, we texted the participants several times a day and asked them how strong a craving to smoke they were having right at that moment, and whether they had smoked since the last time we had texted them. Here's how we made use of those two bits of information. Say you get a text at 2 p.m. and you indicate that you really crave a cigarette right now. When you get another text at 4 p.m. and report that you haven't smoked in the last two hours, it means you didn't give in to the craving and won that particular battle. We could essentially code each battle as to whether self-control or the craving won out.

As expected, the participants were more likely to have smoked by 4 p.m. if they had a strong craving to smoke at 2 p.m. But rVLPFC activity had a major impact on the relationship between craving and smoking. Those with the weakest rVLPFC responses days earlier, during the go/no-go task, tended to go straight from craving to smoking. In contrast, for those with the strongest rVLPFC responses, cravings did not typically lead to smoking between texts. The cravings were still present for these individuals, but the individuals were better equipped to fight the battle. These results imply that the rVLPFC not only is clearly linked to performing self-control tasks in the scanner but plays a significant role in real-world self-control success.

Cognitive self-control. Let me ask you a question. Does the conclusion of the following syllogism logically follow from the premises?

> No addictive things are inexpensive.
> Some cigarettes are inexpensive.
> Therefore, some cigarettes are not addictive.

The question posed to participants is whether the conclusion would have to be true *if* the premises were true. The answer is yes. This conclusion is logically valid, yet fewer than half of the participants answer the question correctly. Why? Because of *belief bias*. We are biased against affirming the conclusion because we know it to be false. The reason it is false is that the first premise of this syllogism is false, but that doesn't make the conclusion logically invalid. Overriding our knowledge of reality in order to imagine a world in which the premises are true requires mental self-control. Although we may not have as much control over our thoughts—our cognitive processes—as we would like, we do have some control, and this control often involves the VLPFC.

To study the neural bases of cognitive self-control, neuroscientists Vinod Goel and Ray Dolan asked people in an MRI to make logic decisions for a series of syllogisms that either invoked or did not invoke the belief bias. They looked at which brain regions were more active when people overcame the belief bias and delivered the correct answer, compared to trials when they did not. The only region of the brain that showed this pattern was the rVLPFC. Another study of the belief bias observed that the strength of activity in the rVLPFC (but not the left VLPFC) predicted how accurately an individual performed. Additionally, being mentally distracted while performing the task led to reduced accuracy and rVLPFC activity. That is consistent with an rVLPFC account of effortful self-control. Finally, a third study used transcranial magnetic stimulation (TMS), described in Chapter 6, to temporarily knock either right or left VLPFC offline for about 20 minutes. Participants were presented with belief bias and nonbelief bias syllogisms both before and after TMS was applied. Individuals who had TMS applied to the rVLPFC, temporarily frazzling the region, performed worse on the belief bias trials. This finding suggests that when the rVLPFC is impaired, self-control is also impaired, leaving individuals less able to overcome their own beliefs to provide the logically correct answers.

A similar finding has been demonstrated with framing effects first discovered by Nobel Laureate Daniel Kahneman and Amos Tversky. Consider the following scenario. Would you rather win $10, no strings attached, or instead flip a coin and win either $20 or nothing at all? Most people go for the sure thing rather than the coin flip. Now imagine a second scenario. This time the experimenter starts by handing you $20. He then gives you the option of giving back $10 to him or flipping a coin to either lose the entire $20 or lose nothing at all. Here, most people go for the coin flip. The strange thing is that the two scenarios are financially identical. In both cases, the sure thing leaves you with $10 more than when you showed up to the study, and the coin flip leaves you with either $20 or $0, relative to when you arrived. People make different choices because the first scenario is framed in terms of winning (that is, winning $10 or having a chance at winning $20), whereas the second scenario is framed in terms of losing (that is, losing $10 or having a chance to lose $20). Psychologically, we are more sensitive to losses so we try to avoid what feels like a sure loss, what Kahneman and Tversky refer to as *loss aversion*.

An fMRI study examined which brain regions were more sensitive to this sort of framing. They found that brain regions in the limbic system were more sensitive to the framing than to the actual facts of the choice. In contrast, the rVLPFC was one of only two regions whose activity was more sensitive to the facts than to the framing of them. As in the belief bias study, rVLPFC activity was associated with overcoming a cognitive impulse.

Perspective taking. In Chapter 5, we focused on the mentalizing system central to mindreading. A lot of mindreading is the same as perspective taking. For instance, in the Sally-Anne false belief task, children succeed to the extent that they appreciate that Sally has a different perspective on things than they do. Typical fMRI studies of mentalizing do not report rVLPFC activity, but a patient, code-named WBA, has provided insight into this region's involvement in perspective taking. WBA had a stroke that selectively damaged his rVLPFC and very little else in the brain. WBA was asked to perform two variants of a false belief task. One of these was easy for WBA, and one was virtually impossible for him.

In one version, WBA watched while a man placed a ball in one of two identical containers, say the one on the left. WBA could also see that there was a woman in the room who was also watching the ball placement. At this point, everyone had seen where the ball had been placed (the container on the left). The woman then exited the room, and while she was gone, the man switched the two containers so the ball was then in the container on the right. When the woman returned, WBA was asked where the woman would look for the ball when prompted. WBA should have pointed to the then empty container on the left because that was where the ball was when the woman last saw it.

In the second version of the task, WBA knew that the man was placing the ball into one of the containers, but he could not see which container it went into. However, the woman could see where the ball was placed, and WBA knew that she could see this. As in the first version, she then exited the room, and while she was gone, the man switched the containers. This time, the experimenters wanted to know if WBA himself could figure out where the ball was. To help him out, the woman who had returned offered to help him out by indicating that she thought the ball was in the container on the right. Even though WBA never saw where the ball was placed, he ought to have inferred that whichever container she pointed to must have been the wrong container because she didn't see the containers being switched. Thus, if she thought that it was in the container on the right, WBA should have chosen the container on the left for himself.

Superficially, these tasks are similar to one another, yet WBA performed magnificently on one and abysmally on the other. Can you guess which was difficult for him? It was the first version—when he could see with his own eyes where the ball was placed, he couldn't do the task at all. In both versions of the task, WBA was aware that the woman had been duped as a result of the container switch while she was out of the room. When that was all that he knew, as in the second version, he had no problems at all using his Theory of Mind to assess what the woman believed. But when he had direct personal knowledge of where the ball actually was, as in the first version, this immediate experience overwhelmed his logical knowledge, leading him to indicate that the woman would look for the ball in the same location that he would. Without his rVLPFC intact, he could not overcome his own first-person perspective and behaved with the same egocentrism of a two-year-old who acts as if everyone sees what he sees and believes what he believes.

We have recently seen something similar in my lab. Imagine being asked the following two questions. First, I offer you $60 to stand in front of Joe's restaurant for an hour, wearing a large sign that says "Eat at Joe's." Would you do it? Second, if I asked lots of people, what percentage of them do you think would say yes? Psychologists have long known that the answer to the first question dramatically biases the answer to the second for most people. If you would wear the sign, you will tend to think most people would. If you would not wear the sign, you will assume most people would also say no. This is called the *false consensus effect* because we tend to believe the world at large shares our beliefs and point of view more than they actually do. Put a different way, we tend to use our own perspective as a proxy for the likely perspective of others. Sometimes this is reasonable, but in many cases this gets us into trouble in social interactions.

To examine the neural bases of the false consensus effect, my graduate student Locke Welborn and I asked UCLA undergraduates lying in an MRI scanner to judge on a scale from 1 to 100 how much the typical UCLA undergraduate would endorse certain positions (for instance, "school prayer" and "abortion rights"). From earlier ratings, we also knew what each participant's own view was on each issue, as well as the actual average response of UCLA undergraduates. With this information, we could tell whether participants' judgments of the typical UCLA student were more in line with reality or were being pulled toward participants' personal views on each issue. As expected, participants did indeed show the false consensus effect, generally judging the typical student to have attitudes closer to their own than they had in reality.

Important to note is that people varied, such that some were better or worse at overcoming this impulse to project their own attitudes onto others. How did people overcome this bias when trying to appreciate the point of view of others? The rVLPFC was one of the only regions of the brain that was more active in participants who were better at resisting their own attitude when considering the typical attitudes of others. The rVLPFC appears to have helped participants appreciate that others might have a perspective different from theirs.

In a sense, both of these studies are like belief bias, only taken into the social domain. We have an immediate intuitive sense of things being a certain way, and it takes self-control to set this perception aside to consider alternative ways of processing the same information. This is an endlessly handy kind of self-control to have in everyday interactions. But on its face, it is so different from the kind of self-control necessary in the go/no-go task that it is difficult to imagine that both kinds depend on the same mental machinery.

Keeping Your Cool

In the summer of 1984, the Gillette Company released a series of advertisements promoting its new antiperspirant Dry Idea. Each of the ads featured someone famous giving a list of three "nevers" central to their line of work. Perhaps the best-known variant was Dan Reeves, NFL coach for the Denver Broncos, describing the three nevers of being a winning coach. In a relaxed pose, he says, "Never let the press pick your starting quarterback. Never take a last-place team lightly. And, really, no matter what the score, never let them see you sweat." His final line of the commercial was, "Everyone feels pressure. Winners don't let it show."

This is the classic image of keeping cool under pressure. You might be making a pitch in the boardroom that you are terrified of blundering, but on the outside you keep your composure as if you have all the confidence in the world. This is a form of emotion regulation that psychologists refer to as *suppression*. This name is a bit misleading because suppression isn't used to suppress one's experience of an emotion but rather to control one's facial expressions, tone of voice, and body language to make sure others can't tell what one is feeling on the inside.

If suppression is the brute force approach to emotion regulation, *reappraisal* is the more cerebral approach. Great thinkers throughout history have commented on our ability to change the way we see things so that they are less distressing. The Roman emperor Marcus Aurelius had a penchant for this strategy, suggesting, "If you are distressed by anything external, the pain is not due to the thing itself but to your estimate of it; and this you have the power to revoke at any moment." My favorite author, Haruki Murakami, condensed this idea to a bumper sticker: "Pain is inevitable. Suffering is optional."

In essence, reappraisal is a process whereby we consider a new perspective that changes how we experience something that is upsetting us. Many reappraisals take the "when God closes a door, he opens a window" approach. You may have gotten fired from a job, but you quickly realize that job wasn't the right job for you anyway. Now you can pursue your lifelong dream of writing jingles for fast-food commercials. From the outside, this realization might seem like rationalizing, just an overly optimistic story that people tell themselves, without changing reality. However, psychologically, our reality derives from the stories we tell ourselves, at least the ones we believe. If you can honestly find ways in which you might be better-off now that you have lost your job, this reappraisal will actually help. Of course, if you believe the job you got fired from was your dream job, it might be hard to believe the reappraisals you come up with.

Personally, I tend to reappraise most when flying. I'm not a big fan of turbulence. When a plane drops five feet suddenly because of an air pocket, my body screams "danger." My heart races, my body starts to sweat, and I search for a window to see if there are any gremlins on the wing. These threat reactions are orchestrated in part by the activation of my amygdala, which is involved in making rapid assessments of the emotional significance of whatever is going on and preparing my mind and body to react swiftly and decisively (though not always intelligently).

I calm my turbulence-induced nerves by thinking of a series of turbulence-relevant facts. First, I think about the fact that my amygdala is not calibrated to make good sense of these quick vertical shifts because such shifts were almost uniformly absent during our evolutionary history—airplanes, elevators, and roller coasters are modern inventions. In other words, I remind myself that even though my amygdala can't make sense of what just happened, I can. Second, I remember the statistics showing that it is incredibly rare for a commercial plane to go down from turbulence. Both of these thoughts help remind me that turbulence and my body's reaction to it are not strong indicators that something is really wrong. The third thing I do, assuming the plane has Wi-Fi, is Google "turbulence reports," which shows a map of all the spots in the U.S. airspace where pilots have reported turbulence today. Simply knowing where the turbulence is, when it will stop, and the fact that all those reports came from pilots who survived the same turbulence I am currently experiencing helps me feel better. Changing how I understand the significance of the turbulence changes how my brain and body respond.

Suppression and reappraisal differ in nearly every way. Suppression is better at making you look like you aren't distressed, whereas reappraisal is better at making you feel less distressed. Suppression is more mentally distracting, and if you engage in suppression during an interaction with someone, it will actually interfere with your memory of the interaction. Reappraisal doesn't cause the same memory deficit. Reappraisal also seems to be employed primarily when you are not in the most intense parts of an emotional reaction. Perhaps it requires a degree of mental clarity to generate good reappraisals, and being emotionally aroused interferes with that process. Suppression and reappraisal also have a different effect on other people in the room. You will probably enjoy being around suppressors less, perhaps because they are giving off fewer emotional signs or maybe because they are preoccupied. Being around suppressors will even increase your heart rate more than being around reappraisers.

Despite these differences between suppression and reappraisal, experientially, cognitively, and socially, they both seem to depend on the VLPFC for their success. For people who reappraise, studies have shown that the VLPFC is activated early on in the emotional episode, and for those who suppress, activity in the VLPFC is turned on later in the emotional episode. But both involve the

VLPFC. In the case of suppression, VLPFC activity is linked to our success at hiding an undesirable facial expression. In reappraisal, VLPFC activity has been linked to diminished amygdala responses and self-reported distress. The longer the time a person spends reappraising, the more the neural activity moves from the left VLPFC over to the right VLPFC, suggesting that the left VLPFC may help initiate the process, while the right VLPFC does more to get the job finished.

Putting Feelings into Words

In each of the forms of self-control we've looked at, there is an experience of applying effort to overcome something. Whether it's withholding a finger press, endorsing a statement known to be untrue, or trying not to lose your temper when your boss yells at you, there is an urge or impulse that you can feel yourself fighting against. Yet sometimes the same mechanisms of self-control can be engaged without our even knowing it. The author Henry Miller once wrote, "The best way to get over a woman is to turn her into literature." Putting our feelings into words can be tremendously cathartic and is the basis for various psychological therapies. But it turns out that putting our feelings into words or simply being able to label them can regulate our emotions and promote our mental and physical well-being without our realizing it at all.

When young children are emotional, we tell them to "use their words." Preschoolers who can describe their feelings have fewer emotional outbursts, get better grades, and are more popular with their peers. High school students who write about their math anxiety right before taking a math test actually do better on the test. In my lab, we ask adults to perform a simple task, called *affect labeling*, during which people choose a word to best describe the emotional aspect of a picture. For instance, a picture of an angry face might be shown, and the participant would have to choose whether the word *angry* or *scared* describes the target's emotion. We have found that labeling the affective aspect of a disturbing image reduces the distress a person feels while looking at the image. Even though this result looks like what we might expect from an emotion regulation strategy such as reappraisal, people do not realize that affect labeling is an effective strategy for diminishing their negative feelings. To examine people's beliefs about affect labeling, we have asked individuals to predict which would be more distressing, to look at a disturbing image with no instruction or to look at the image and label the emotional aspect of it. People almost always predict that labeling would be worse because it would focus their attention on the upsetting parts of the image.

To get a sense of how counterintuitive affect labeling effects are, imagine you have a severe fear of spiders and you have gone to get treatment for your phobia. The therapist is going to put you through one of three kinds of treatment regimens. She describes the three versions and lets you choose. The first is a standard type of *exposure therapy,* which involves repeatedly seeing a real tarantula two feet away in its cage. The second is a reappraisal treatment, which also has repeated exposures to a real tarantula, but each time the spider is presented, you will be asked to generate a reappraisal such as "Looking at the little spider isn't actually dangerous for me." The third option is an affect labeling treatment, which again involves the repeated exposures, but this time while generating affect label–based statements such as "I feel anxious that the disgusting tarantula will jump on me." Which kind of therapy do you think would help you learn to approach the spider with less fear? Katharina Kircanski, Michelle Craske, and I ran exactly this test with spider phobics, and we found that affect labeling helped the most and that the more negative the participants' labels were, the better the final results.

Just like reappraisal, affect labeling regulates our emotions and thus appears to be a kind of *implicit self-control.* Does this kind of self-control look like the others in terms of brain activity? Absolutely. When people label an emotional picture or their own emotional response to a picture, it activates the rVLPFC and reduces activity in the amygdala. We've run a number of studies now in which the same individuals use affect labeling, reappraisal, and in one case a motor self-control task. We have seen similar things going on in the rVLPFC across these different forms of self-control.

And thus ends our tour of self-control variations. Whether we exercise self-control over our motor and visceral impulses, logical reasoning, social perspective taking, or emotion regulation, the rVLPFC almost always seems to be at the center of the action. It is still unclear what exactly the rVLPFC does to stimulate self-control. The debate typically focuses on whether this and similar regions are directly inhibiting responses in other brain regions, like the amygdala, or are helping to strengthen the nonimpulsive alternative so it can compete effectively with the impulsive response. In any event, the question I want to turn to now is why self-control plays a central role in our tendency toward social harmony.

Alien Abductions

So far it appears that self-control is a tremendous asset, and when we use it, it involves the rVLPFC region of the brain. Its relation to our sociality emerges when we begin to deconstruct the meaning of the word. The word *self-control* yields two very different meanings, two ways in which *self* and *control* are related to one another. On the one hand, the hyphenated word can imply that our self is in control, achieving its own ends effectively. This interpretation brings to mind the notion of *willpower*, a muscular Nietzschean word for our ability to overcome whatever gets in our way through sheer personal force of mind. But there is a second, more Orwellian connotation, linking self-control with *self-restraint*. Here, it is the self that is being controlled, which leads to the question "Who benefits when we bring the self under control?"

Perhaps a couple of hypothetical alien abductions will help us get to the bottom of this. Imagine that while you are sleeping soundly, little green men snatch you from your warm bed and take you to their advanced neurosurgery facility in the sky. They are deciding whether to alter your brain such that you permanently lose all impulses, urges, desires, and emotional reactions or to leave those intact and instead perform a surgery that will leave you permanently unable to control your impulses, urges, desires, and emotional reactions. The aliens cannot decide among themselves so they let you cast the tiebreaking vote. Which would you prefer to lose if you had to lose either emotion or self-control forever? It's the classic battle between self-control and emotion, between Mr. Spock and Captain Kirk, between businessman and Burning Man.

After multiple failed escape attempts, I suspect that I would ultimately choose to keep my impulses, urges, desires, and emotions and give up my ability to control any of them. It would be embarrassing to lack self-control, but it would be devastating to lose the rest. Who am I without all of these? How would I know what is worth doing? Without impulses and emotions, I would have no motivation to do anything. Remember that not all impulses and urges are bad. I have the urge to kiss my wife and son every day. I have impulses to help those in need. I have the desire to hike up mountains and watch the sun set. These are all wonderful things without which I am not sure life would be worth living.

Unfortunately, even though you have made your choice, things get more complicated. Before performing their operation on you, the aliens suddenly perfect a new technology that allows them to perform neurosurgery on all the inhabitants of a city at once, while they sleep in their beds. They are going to start with your city, but because you are onboard their spaceship, you are now exempt from the surgery. You personally get a reprieve; you will keep both your emotions and your ability to control them. However, you now have to choose whether all the people in your city will lose their ability to feel their impulses and emotions or will lose their capacity for self-control. Whatever you decide will be applied to everyone, so you will be returning to either a city full of highly impulsive, emotional people or a city full of nonimpulsive, highly controlled people. An added note: your decision will not affect your family or close friends because luckily for them, they were all away on vacation.

What do you choose for all the people who make up your city (but are not part of your immediate social networks)? Do you want to live in Kirkville or Spocktown? For me, and I suspect for many of you, this decision yields a different result from what I wanted for myself. I don't want to live in a city full of people who are impulsive nonstop without the ability to control themselves. These people will be reckless and a constant threat to my safety. It would be like living next door to a fraternity house where it's always 1 a.m. on Saturday morning.

These two hypothetical decisions suggest that I value other people's having self-control more than I value having it myself. Assuming this preference is generally true, we can turn it around. If I value other people's having self-control more than I value having it myself, it follows that the people around me care more about my having and exercising self-control than I do. My self-control is more of a benefit to them than to me.

Who Benefits from Self-Control?

The novel *A Single Man* by Christopher Isherwood opens with the main character's morning routine. As he wakes, George is merely "experience—an entity experiencing" without any self-awareness. There are impulses, urges, and even aches. Pure experiences. But then he looks in the mirror. "It stares and stares . . . until the cortex orders it impatiently to wash, to shave, to brush its hair. Its nakedness has to be covered. . . . Its behavior must be acceptable to them. . . . Obediently, it washes, shaves, and brushes its hair; for it accepts its responsibilities to the others. It is even glad that it has a place among them. It knows what is expected of it" (p. 11).

Self-control is the price of admission to society. If you don't restrain your impulses, you will end up in prison or a psych ward. If you do restrain your impulses, you are allowed to freely pursue your goals. And there are nonpunitive incentives for self-control as well. People with greater self-control get paid more because self-control allows those individuals to do things that are of great value to the rest of society. The thing is, just as with the alien abduction scenarios, society values our self-control more than it values our quality of life. John Lennon once told a story about his early education that underscores this. He said, "When I went to school, they asked me what I wanted to be when I grow up. I wrote down 'happy.' They told me I didn't understand the assignment, and I told them they didn't understand life." For his teachers, what he wanted to be had to reflect what he would do that would benefit society. His happiness was a nonsensical answer to them.

How many people devote countless hours of effort, requiring deep reservoirs of self-control, in order to get into medical school, where even greater self-control is required to make it through the internships and residencies, only to find out that being a doctor does not make them terribly happy. Fewer than half of the doctors in the United States say they would choose the same career if they had it to do over. The world respects doctors because they do something that provides a profound benefit to the rest of us. Adolescents want to be respected and wealthy, they want to make their parents proud, but all of the self-control that doctors-to-be apply in the pursuit of becoming a doctor might ultimately be more valuable to *us* than it is to *them*.

Pursuing careers that benefit others more than oneself might be an accidental confluence of factors, but it is not uncommon for societal norms to push people to engage in self-restraint in order to benefit the greater good. In Beijing, many men across a wide range of ages and classes engage in a behavior that has earned them the name *bang ye*, which literally means "exposing grandfathers." These men roll their shirts up above their bellies on the hottest days of the year. In recent years, Beijing has been working to become a cosmopolitan destination city, and these bare-midriffed men conflict with this image. Both the government and the newspapers have run campaigns to try to put a stop to this. This is a case in which self-control would clearly benefit society but not the individuals. Rolling up their shirts keeps the men cooler, but the society at large feels more in line with Cicero's directive that "every man should bear his own grievances rather than detract from the comforts of others."

As individuals and as a society, we have greater general trust in those who display self-control. Both with strangers and with our romantic partners, studies have demonstrated that signs of self-control warrant greater trust. This makes good sense in the case of romantic partners, for those low in self-control report having greater difficulty in staying faithful.

Society bestows some of its greatest rewards to those with high self-control: admissions to top universities and scholarships to pay for them. We have already seen that the major determinants of admission, a student's GPA and SAT scores, are both highly influenced by self-control. We think of the SAT as an intelligence test and thus think of admissions to top universities as an intelligence competition. Though there is truth to this, admission to top schools is just as much a competition over self-control. How much were you able to restrain all of your distracting impulses through thirteen years of school and in studying for the SAT? We might endorse the SAT as the ticket to admission, believing it to separate the smartest from the rest. Indeed, the creators of the SAT designed it to be a measure of intelligence that could not be gamed through practice or hard work. But ultimately, we as a society give people access to top universities based on a test that can be conquered through self-control.

The reason why self-control, when considered across the wide spectrum of behavior, benefits society more than individuals comes down to different cost-benefit equations for individuals versus society. Say you are a smoker and you want to give it up. Even though you know in the long run quitting is far better for your health, it is very difficult to succeed in quitting. Why? Because the short-term benefits of smoking compete with the long-term benefits of not smoking. This may sound sacrilegious, but if you are addicted to nicotine, then it is truly in your immediate self-interest to have a cigarette right now because having one feels far better than not having one. It literally pains the body not to have a cigarette when cravings come on strong. It is only because you can focus on the long-term benefits of not smoking that you may be able to stave off the urge to smoke. For the individual, there is a short-term benefit to smoking and a long-term benefit to not smoking, and the individual must wage a battle between the two.

For society, there is no such trade-off. Society gets almost no short-term benefit from your smoking. It cannot enjoy the cool flavor, experience the nicotine rush, or feel your nerves calming down. For society, your smoking is bad from nearly all angles, at all times, and your not smoking is good from almost all angles at all times. When we think of self-control as willpower, it conjures up visions of the rugged power of the individual overcoming any obstacle. But when we think of self-control as self-restraint, it leads us to wonder whether the individual is actually the prime beneficiary of those self-control efforts. Self-control typically pits momentary happiness against an abstract better life in the future. That abstract better life almost always aligns with society's goals, but as John Lennon implied, your momentary happiness is not a priority for society.

As we have said from the beginning, we think people are built to maximize their own pleasure and minimize their own pain. In reality, we are actually built to overcome our own pleasure and increase our own pain in the service of following society's norms. Once again, this highlights how poor our theory of "who we are" is. Yet the first textbook ever written on social psychology, by Floyd Allport, nailed this idea almost a century ago. Allport argued that "socialized behavior is thus the supreme achievement of the cortex. . . . It establishes habits of response in the individual for social as well as for individual ends, inhibiting and modifying primitive self-seeking reflexes into activities which adjust the individual to the social as well as to the non-social environment."

We discussed in the last chapter how the medial prefrontal cortex (MPFC) may serve as a Trojan horse for social influence, allowing the beliefs and values of whatever society we mature in to become internalized and treated as personal beliefs and values, without our realizing that this psychological invasion has taken place. Despite these beliefs and values becoming something we strongly endorse, they sometimes have difficulty competing with our unsocialized urges and impulses. As comedian Louis C.K. once said, "I have a lot of beliefs, and I live by none of them."

Having the same beliefs as others in our group (for example, classroom, business, society) helps us harmonize—to get along and like one another. Most of the time we can get by assuring others we share their beliefs and values without having to enact them. The MPFC may make sure we talk the talk, but some insurance was needed to make sure we walk the walk, and that is where the VLPFC comes in. If we are sufficiently motivated, the VLPFC can help us to tip the balance so that our thoughts, feelings, *and* behaviors are guided by our socialized beliefs and values over the strong dictates of our presocialized urges and impulses.

Who Controls Our Self-Control?

In the eighteenth century, British philosopher Jeremy Bentham proposed something that would lead to "morals reformed—health preserved—industry invigorated." He had designed a new kind of building called a *panopticon* that he believed was the key to making people do the things they should. Bentham's plan was to make all of the people in a particular group, whether prisoners, students, employees, or hospital patients, able to be observed at all times long before security cameras would achieve this end. The plan was to build rooms around the perimeter of a circle facing in toward the open space in the middle. In the case of a prison, an individual cell would have solid walls on each of its sides except for the side facing the middle of the circle—that side would have bars keeping the individual in, but it would be open otherwise. In the middle of the circle, a guard tower would sit with 360-degree views of all of the cells, which could be stacked several floors high. This would allow a single prison guard, "without so much as a change of posture, [to view] half of the whole number" of prisoners.

The remaining architectural element is what made the panopticon ingenious. Bentham suggested that, ideally, each prisoner would be watched by a guard at all times because being watched and the threat of punishment would keep prisoners in line. The central tower would give the greatest viewing span possible, yet still it wasn't possible to have a single guard or even a few guards truly pay attention to all prisoners at all times. Bentham's solution was to ensure that all prisoners felt they *might* be under a watchful eye at any given moment without knowing whether they actually were or were not. The guard tower was to be built so that guards could see out but prisoners could not see in. A prisoner might be watched at any time "and not being able to satisfy himself to the contrary, he should *conceive* himself to be so."

Bentham's great insight was that being watched by an authority is not the only way to keep people in line. Rather, it is sufficient for people to know they are *seeable*, that there is the potential for them to be judged, evaluated, and punished. Being seeable is enough to trigger us to restrain our unsocialized impulses through self-control.

Evolution has wired us with *panoptic self-control* in which the mere possibility of being judged and evaluated by others dramatically increases our tendency to behave in line with society's values and morals. This might seem like a strategic, rational thing to do. If I'm going to steal, I should steal only when I am sure I cannot be caught. Until then, I should behave publicly like someone who has no intention of stealing. But as we will see, research shows that people take this to the extreme, beyond what is rational.

In one study, an experimenter "accidentally" dropped a large stack of papers in front of the participant. Sometimes there was a visible security camera in the room, and sometimes there wasn't. The subject was 30 percent more likely to assist the experimenter if the security camera was present, indicating the possibility of additional observers. Another study found that individuals were twice as likely to cheat on a test in a room that had the lights dimmed, making observation of their immoral behavior more difficult. In a third study, subjects wearing eye tracking devices (that allowed the experimenter to reconstruct where the subjects had been looking) were much less likely to look at a risqué poster on the wall than those not wearing this device. In each of these cases, the subjects were calibrating their behavior to reality. The more likely they were to be observed, the more they acted like it.

Panoptic self-control also goes beyond these rational responses to situations in which there is merely an abstract indicator of being seeable, even when we know we are not actually being observed. Imagine in the break room at work, there is an "honesty box" where you are meant to pay for whatever drink you take from the fridge. There is a poster on the wall indicating the price of each drink; you are meant to drop that amount in the honesty box. No one else is in the room with you, and you don't hear any footsteps coming toward you. Do you pay for your drink? If so, how much? Would it matter if the poster on the wall had some pictures of flowers on it? How about if the poster had a picture of a pair of eyes on it? Not real eyes, not a security camera—just a photograph of someone's eyes, eyes that can in no way actually see what you are doing. Compared to the flower-adorned poster, the one with the eyes led people to pay 276 percent more into the honesty box. In a public cafeteria, a similar "eyes poster" nearly halved the amount of littering that people did. Even pictures of a defunct toy robot's eyes led to larger donations in a laboratory-based economics game. Finally, my favorite: a triangle of three dots in the approximate configuration of two eyes and a mouth contrasted with three dots positioned so that the single dot is at the top (see Figure 9.2). Men presented with the "face" version were three times as likely to donate money to another player in an economics game as men seeing the triangle with the single dot at the top.

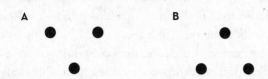

Figure 9.2 Dot Configurations That (A) Do and (B) Do Not Induce Prosocial Behavior. Adapted from Rigdon, M., et al. (2009). Minimal social cues in the dictator game. *Journal of Economic Psychology,* 30(3), 358–367.

It is one thing to strategically take into account whether you are being seen before you engage in bad behavior, but what does a photograph of eyes or dots forming a triangle imply about the actual likelihood of your getting caught and punished? Rationally, people in these studies can tell you that they know they are not being watched and are not likely to be caught, no matter what they choose to do. Nevertheless, people restrain themselves *as if* they might be seen.

Panopticons of the Mind

Think back to the days of your youth when October 31 represented the single best opportunity of the year to gorge yourself with so much candy you might actually regret it afterward. On Halloween, all you have to do is knock on a stranger's front door in some semblance of a costume, and you are rewarded with candy. Imagine you walk up to the forty-second front door of the evening, and just after the owner of the home greets you, he gets an important phone call. He says, "I'm sorry, but I need to take this call. The candy bowl is right inside the door. Go ahead and take one piece of candy. But I need to go to the other room." He walks away and leaves you alone with a large bowl of candy. What do you do? Do you take a single piece as he invited you to do, or do you put as much in your bag as you can, as quickly as possible. No one can see you. Well, one person can see you—you. Behind the candy is a mirror that reflects your own actions back to you. Would this affect your decision?

Apparently, when put in this situation, our natural impulse is to take more than we should. When children (ages nine and up) were put in this scenario without a mirror, a little more than half of them took more than the one piece of candy that they had been instructed to take. But when they could see themselves in the mirror, fewer than 10 percent of the children took more than one piece of candy. This is staggering. The mirror made children five times less likely to violate a social norm. Just seeing one's own reflection is enough to bring our self-control online to overcome the impulse to snag some extra candy.

A century ago, George Herbert Mead and Charles Cooley each suggested that self-consciousness is essentially a dialogue between our impulsive self and a simulation of what we imagine people important to us would say to us if they knew what our impulsive self was getting ready to do. We experience self-consciousness as a private internal process, but according to these psychologists, it is actually a highly social process during which we are reminded of what society expects of us and then we prod ourselves accordingly. In essence, this view suggests that we are our own panopticon: both seer and seen.

This isn't just about Halloween trick-or-treatery in young children though. In a laboratory study, first-year college students were ten times less likely to cheat on a test in the presence of a mirror (71 versus 7 percent). In the absence of any observers, the natural impulse is to cheat (apparently), but people uniformly restrain this impulse when they see themselves. People are also more likely to conform to others in the presence of mirrors across a variety of contexts.

Other species exhibit self-control, and some can even recognize themselves in the mirror, but only humans are built such that seeing themselves, a reminder of their potential visibility to others, is sufficient to trigger self-restraint. Seeing ourselves as others would see us (that is, our visible appearance) is sufficient to engage our self-control to overcome our unsocialized impulses in order to fall in line with society's expectations. When we began our discussion of self-control, it seemed like a mechanism that would primarily support our own individual interests, putting ourselves in control of our lives. As we have seen, self-control operates at least as often to benefit society. We are built such that the most trivial reminders of ourselves as social objects keep us in check. Self-control enhances social connection because it helps us to prioritize the good of the group over our own narrow self-interest. Self-control increases our value to the social group, and by conforming to group norms, we reinforce the group's identity as well. Self-control is a source of social cohesion within the group, putting the group before the individual. This is the essence of *harmonizing*.

Reminders that we are the kind of creatures that can be seen, judged, and evaluated engage our self-restraint in the service of prosocial outcomes like not cheating and conforming to group norms. These three processes (being evaluated, engaging in self-restraint, and complying with social norms) seem quite distinct from one another taken at face value, and yet there is reason to think they are each tied to rVLPFC functioning, efficiently converting our sense of being judged by others into self-control efforts that result in social compliance. We have already encountered plenty of evidence on the role of the rVLPFC in self-restraint, so let's focus on the other two processes.

Imagine an experimenter gives you $100 and asks you to decide how much you want to give to another person in the experiment.

You won't ever meet that person, but he is real, sitting in the next room, and he knows you've been given the money to split between the two of you. You have complete control over the money. How much do you give? What are the options that go through your mind? Given that neither of you earned the money, the social norm of fairness prescribes that you should split it down the middle, fifty-fifty. However, selfish motives dictate taking as much as you can get away with. Under these conditions, people tend to give around 10 percent when they know there will be no further interactions with the other person.

Manfred Spitzer and Ernst Fehr ran this study in the scanner along with another condition in which participants felt pressure to comply with the social norm. Imagine that the person in the other room could punish you after finding out how you had decided to split the money. He could use some of his own money to dramatically decrease the amount of money you ended up with; for every dollar he spends, you would lose $5 that you had allocated to yourself. Knowing that he could punish you, how much would you give to him now? In this condition, people came much closer to the fifty-fifty norm, giving about 40 percent of the money to the other person.

When people complied with the fairness norm in this study, it is not as if they wanted to give that much money. If they truly wanted to, then they would have given 40 percent in the control condition as well. Instead, they do it because they feel pressured to act fairly. The rVLPFC was one of a handful of regions that was more active during the social compliance trials. Of course, this region might be sensitive to the threat of losing money, rather than to the pressure to comply with a social norm. To address this, Spitzer and Fehr compared their results to other conditions during which the participants played with a computer rather than with a real person. The threat of punishment produced more right VLPFC activity when the threat came from a real person, even though the financial dynamics were the same in both cases. In other words, this region seems to be involved in converting the threat of social sanctions into compliance with social norms. As it turns out, other studies have shown that merely seeing another person rate something highly, like a song, can move us to rate that thing more highly too. People who conform most to this kind of norm set by others show increased activity in the rVLPFC and actually have more gray matter in this region.

These compliance studies focus on situations in which our initial plans or evaluations differ from those of others around us. The notion of panoptic self-control suggests that the mere possibility of being socially evaluated is sufficient to engage self-restraint. Although no one has looked at this directly, there is research indicating that just imagining what others think of you is sufficient to activate this rVLPFC region. The most striking finding on panoptic self-control is that merely seeing yourself, with no one else around, can promote self-restraint as well. Can you guess which brain region is most consistently activated when you see a picture of your own face? Yes, the rVLPFC. When you see a picture of yourself, reminding you of how you look to the outside world, you turn on the same part of your brain that is responsible for self-restraint and for compliance with social norms. The connection between these three functions within the rVLPFC has not been systematically investigated yet, so its true significance is still a mystery. However, one intriguing possibility is that these processes became linked over the course of evolution in order to ensure that we would use our fear of not fitting in socially to engage our capacity to override our more indulgent self-interests.

That is about as far as one could get from the version of self-control we started out with. Our intuitive notion of the role of self-control is to promote our individual private goals and values. This new evidence suggests it is more of a mechanism to help shape our behavior to be in line with the group's goals and values when they conflict with our own. We tend to think of people who conform as lacking in courage and initiative—as weak-willed sheep following the herd. Yet the current analysis suggests that in certain situations, people with the greatest capacity for self-control will actually conform more than other people. Sometimes the threat or perceived threat of sanctions from the group makes conformity the smart choice, and those with more self-control will be better able to overcome the urge to act impulsively.

The Self Is For?

In the West, our conception of the self is as a treasure trove of thoughts, feelings, and desires that represent who we really are. To "know thyself" allows us to expend our limited resources seeking out and working toward the things that will truly make us happy and to avoid the things that will make us unhappy, whether in the short term or the long term. And this account definitely holds some water. It's useful for me to know what kinds of foods I like, what kinds of social events make me uncomfortable, and which kinds of work will help me feel most fulfilled. Having a theory of my own mind is an eminently useful thing.

What we fail to appreciate, however, is the degree to which society has shaped the contents of our minds—the way we form our goals and beliefs, and what causes us to exert our self-control in different situations. From infancy, we are surrounded by a social world that is more than happy to tell us what good people want and do, to tell us which of these desirable characteristics we have, and what kind of life is worth leading. However, all of this input from the outside world would amount to nothing if we weren't born with a Trojan horse self that is built to soak all of this up like a sponge, without us realizing where these foundational worldviews came from. We believe these are our deeply personal private stock of beliefs, and that notion makes us work hard to defend them. It rarely dawns on us that others put them there. When we defend our beliefs, we are usually defending society's beliefs. This alignment between our private beliefs and the beliefs of those around us motivates us to be useful members of society. It helps to ensure that others will like us, and it increases the ratio of social pleasure to pain we will encounter in our lives.

Self-control to us feels like a source of power—the willpower that allows us to drive our personal agenda forward. It may be easily depleted, but it has the unique capacity of overriding our momentary desires in order to implement our personal long-term goals. But as we have seen, our personal, long-term goals nearly always benefit society as much as or more than they benefit ourselves. And when there is a conflict between our personal values and those of society, simply being reminded that we can be seen and judged by others activates our panoptic self-control to override our impulses, bringing our behavior in line with societal expectations.

These are very counterintuitive notions. The idea that our personal values were snuck into us by society at large and that our self-control exists in part to restrain, rather than support, the self is anathema to our way of thinking about "who we are." Yet brain science is helping us to see the fundamental truth behind these claims—that our most deeply personal sense of self and sources of willpower may most often serve to keep us in the good graces of the group. Harmonizing is hard work, but apparently evolution thought it was "worth it" to make our attitudes and beliefs aligned with those of the group rather than at odds with them.

Our Social Brain

And that's the story, folks, at least the neuroscience part of the story. Over the course of millions of years of evolution, our brains have marched ever increasingly to the beat of a social drum. To have larger brains that could solve all manner of problems, evolution had to first solve the problem of getting those brains out of the womb. The solution was immature brains that would do most of their growing outside in the light. This necessitated *connection* as a central mammalian adaptation so that mammalian young would be cared for during infancy and stick around to do the caring in adulthood. We have seen that this necessity was implemented through dual mechanisms. Social pain, via the dorsal anterior cingulate cortex and the anterior insula, sounds an alarm that motivates us to address threats to our connections. Social rewards, via the ventral striatum, the septal area, and oxytocin processes, all play a role in the pleasure we take from feeling cared for and motivate us to care for others.

As primates emerged on the scene, the rudiments of *mindreading* evolved. Mirror neurons in lateral frontoparietal areas allow us to imitate and thus learn from the actions of others. Critically, these regions also appear to be central to humans, representing actions as actions, replete with psychological meaning. With the emergence of the mentalizing system in humans, in the dorsomedial prefrontal cortex and the temporoparietal junction, we are uniquely capable of reasoning about the actions of others and what that emergent ability tells us about their thoughts, feelings, and goals. This capacity is so important that over the course of evolution it was selected for such that the mentalizing system comes on spontaneously whenever there is no other mental task occupying us. This resetting nudges us to see the world in terms of its social and mental elements rather than its physical elements. Indeed, this resetting involves the brain muting its own circuitry for nonsocial reasoning. Mindreading is instrumental in rationally pursuing our social motivations: finding ways to enhance our social connections and avoiding the pain of social rejection.

The coup de grâce of evolution's molding of a social brain is the twin stars of self-knowledge and self-control. Our sense of self, as represented in the medial prefrontal cortex, is largely a deceit. What it contains we believe to be private and inaccessible, yet in reality it is a conduit for the socialization of our beliefs and values. Self-control, as mediated by the ventrolateral prefrontal cortex, also serves a purpose different from the one we first imagine. Rather than pushing our own personal destiny forward, self-control often serves as an instrument of social control ensuring that we follow social norms and values. In a sense, neither the self nor self-control end up serving us in the way we imagine they should. They do, however, serve to ensure social *harmonizing*. They make us more likeable and agreeable to others in the groups we spend our time with. They make us strive to support the group, sometimes at the expense of our private unsocialized impulses, and this effort makes us more valuable to the group. All of us operating with the same tendency to prioritize the group allows the group to thrive in the face of competing private interests that are ever present as well.

Living social lives is difficult, really difficult. We depend on the most complicated entities in the universe, other people, to make our food, pay our rent, and provide for our general well-being. This system is far from perfect, but evolution has bet time and time again on making us more social.

Part Five

Smarter, Happier,
More Productive

CHAPTER 10

Living with a Social Brain

The message is clear: our brain is profoundly social, with some of the oldest social wiring dating back more than 100 million years. Our wiring motivates us to stay connected. It returns our attention again and again to understanding the minds of the people around us like a rubber band snapping back into place. And we have this center to our being, what we call our self, which among its many jobs serves to ensure that we harmonize with those around us by lining up our beliefs with theirs and nudging us to control our impulses for the good of the group. The biological depth of our sociality is important because it fleshes out a woefully incomplete theory most of us have about "who we are."

We look around us and see people selfishly motivated by pleasure and pain and little else. This is what we've been taught for generations, and it is true that these are powerful motivators of human behavior, but they are far from the whole story. If we keep eyes open for it, we will see plenty of behaviors that we can't quite square with self-interest as the sole motivator in our lives. We have failed to understand them because we have failed to fully understand what kind of beings we are.

So where do we go from here? Is understanding our social brain merely an intellectual exercise, satisfying an existential urge to know what it means to be us? While that's an itch I'm always happy to have scratched, I think understanding the nature of our sociality is far more significant than that. Everything we do in life and all the organizations that we are a part of are affected by our understanding of "who we are."

Think about how amazing the brain is, and then consider that a huge portion of that amazing brain helps to make us more social. Yet for a large part of our day, whether we are at work or at school, this extraordinary social machinery in our heads is viewed as a distraction, something that can only get us into trouble and take us away from focusing effectively on the "real" task at hand.

Chapters 10, 11, and 12 reveal how wrong this view is. Almost everything in life can be better when we get more social. If we retune our institutions and our own goals just a bit, we can be smarter, happier, and more productive.

The Price of Happiness

We all want a good life—to be happy and healthy. Society at large has a huge investment in people being happy and healthy as well; happy and healthy people are more productive, get into less trouble, and cost society less money. Philosopher Jeremy Bentham founded the Utilitarian school of thought on the notion of the *greatest happiness principle*, or the idea that the best society has the greatest amount of pleasure relative to its pains. The big question—a question that has been asked for as long as we have been asking questions—is what makes for a happy and healthy life. If we have been getting this wrong, we should all want to know so we can start getting it more right.

In 1989, more than 200,000 college freshmen were asked about their life goals, and one goal stood out from the rest—to be well-off financially. Perhaps these students had been reading Ayn Rand's novel *Atlas Shrugged,* in which one of the characters declares that "money is the root of all good." Or perhaps they were just being sensible. If Bentham is right about pains and pleasures, having piles of money is a great way to avoid physical discomforts and to maximize access to life's material pleasures. Want to travel to exotic places and eat the world's finest cuisines? Want to go farther and orbit Earth from space? You can do all of these things, but only if you've got the bank account to get you there. There's no question that making more money is valued worldwide and that it provides access to countless resources. But does it make us happy?

Economists have been obsessed with this question for several decades, in part because the income of individuals and nations was long taken as an objective indicator of their well-being, for it was believed that true well-being could not be directly measured. This assumption in some ways may have led money to be seen by society as an end in itself, rather than as a means to an end. Despite the presumption that true well-being cannot be directly measured, "happiness," "life satisfaction," and "subjective well-being" are actually measured quite easily. All you have to do is ask people questions like "All things considered, how satisfied are you with your life as a whole these days?" and people will tell you. If you ask the same people today and a year from now, you will pretty much get the same answer from folks both times. People have stable, reliable responses to this kind of question.

There are many ways to tackle the relationship between money and well-being, and economists seem to have tried them all. The surprising conclusion of nearly every approach is that money has much less to do with happiness than we think it does. Let's begin with the one and only analysis that ever shows strong links between money and well-being. If we look at a large number of countries and for each nation get an average measure of well-being and get the nation's average income level, these two factors will be correlated quite highly. Countries with higher average income have citizens who report higher average well-being. But this kind of analysis may not tell us very much because rich countries differ from poor countries in countless other ways. Rich countries allow for more individual freedom, have better schools and health care, and have less corrupt judicial systems. Gross domestic product may just be a proxy for one or more of these other variables that might affect happiness more directly.

Let's consider some of the other tests. Researchers have also looked at the link between money and well-being within particular countries. For instance, happiness researcher Ed Diener looked at surveys of thousands of U.S. adults who reported their subjective well-being and their income. There was a statistically significant relationship between how much a person earned and how happy they were, but it was extremely modest. Individuals' income explained only about 2 percent of the differences in happiness across the sample. And most of this relationship has to do with being below or above the poverty line. If you are below the poverty line, every additional $1,000 you earn dramatically alters your well-being. But once the basic needs are met, increasing income only adds the tiniest bit to well-being.

Some have suggested that the proper way to isolate the relationship between income and well-being is to look for changes in income over time. For instance, one study examined changing U.S. income levels between 1946 and 1990 and compared these to changing self-reports of well-being. The results, shown in Figure 10.1, are striking. Income, after controlling for inflation, more than doubled during this time, and yet well-being did not increase at all. This effect, called the *Easterlin Paradox* for the economist who first discovered it, has been shown for many countries, but for none more dramatically than Japan. Between 1958 and 1987, real income increased 500 percent and material comforts multiplied similarly (for example, car ownership grew from 1 percent to 60 percent). Nevertheless, Japanese reported equal levels of well-being across these three decades.

I don't know about you, but I find this all very disconcerting. I work hard for my money, and I work hard to make more of it. I do this because I know in my gut that if I can make more, my family and I will be happier. This brings us to the last stop on the money train to happiness. Some economists have tracked individuals across a decade or so to see if changes in their personal income level are associated with concomitant changes in well-being. They aren't.

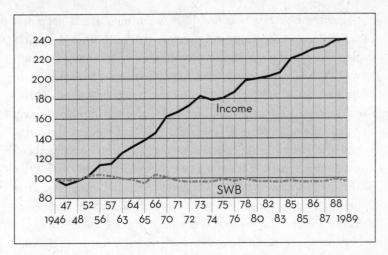

Figure 10.1 Changes in U.S. Income and Social Well-Being (1946 to 1989).
Adapted from Easterlin, R. A. (1995). Will raising the incomes of all increase the happiness
of all? *Journal of Economic Behavior & Organization,* 27(1), 35–47.

Some people were making substantially more money at the end of
ten years and some were making substantially less, but well-being
was unrelated to these changes. My gut says making more money
will make me happier, but my gut is wrong.

Explaining the Paradox

I shouldn't be alone in my frustration at this point. The vast majority of people indicate that making more money is one of their primary life goals. We don't do this for the heck of it. We do it because
we believe it will give us a better life in the end. Yet study after study
reaches the same conclusion: it won't. We have been barking up the
wrong tree. How could we, as a society, have gotten this so wrong
for so long? What are we missing to end up so misguided in our
theory about what will make us happy?

After the lack of a relationship between money and happiness came to light, economists and psychologists each offered sensible explanations of the missing relationship between money and happiness. Psychologists pointed out that humans have the tendency to adapt to new circumstances, whether they are good or bad. This is called *hedonic adaptation,* and in many situations, it helps protect us from staying depressed forever over negative events. Unfortunately, the same mental machinery keeps us from staying elated after positive events. The most famous example of this is the case of the major lottery winners who were contacted some time after winning. They reported being no happier than individuals from the same communities who had not won.

Economists generated a second explanation focusing on the context in which one's income is considered. They suggested that the problem is that we focus less on our absolute income and purchasing power and more on how much we are making relative to those around us. This *relative income* argument suggests that earning $50,000 a year in a neighborhood where most people earn $30,000 a year could make us happier than earning $100,000 a year and having neighbors who earn $200,000 a year.

Missing Social

Things are even worse than I've portrayed them, at least in the United States. Not only is increasing income not associated with increased well-being over the past several decades—but well-being has actually decreased over this time period. Sensitivity to relative income definitely accounts for part of this reduction in happiness, but it's not the whole picture. Something else is going on that can't be explained by these kinds of factors. In the book *Bowling Alone,* Robert Putman first put his finger on what was missing in all of these analyses: social. Putnam and those who have followed him have put together a number of variations on the same two-step theme. First, social factors substantially contribute to subjective well-being and life satisfaction. Second, in modern nations like the United States, these social factors are in decline. Let's take these in order.

Economists use terms that sound like economic indicators, such as *social capital* and *relational goods*, to talk about a variety of social factors. These include being married, having friends, the size of one's social network, whether people join social organizations (such as bowling leagues), and trust in various societal institutions. Pretty much any way economists examine these social factors, they (unlike income) end up being significantly related to well-being. One study compared the impact of income and social connections on well-being and found that social factors had a more positive impact on well-being than income, once relative income effects were considered.

Just how much are the social aspects of our lives worth in terms of our well-being? Multiple studies have managed to put a dollar value on them, determining how much more money you would need to make in order to achieve the same increases in well-being. In one study, volunteering was associated with greater well-being, and for people who volunteered at least once a week, the increase in their well-being was equivalent to the increase associated with moving from a $20,000-a-year salary to a $75,000-a-year salary. A second study found that across more than 100 countries, giving to charity is related to changes in well-being equivalent to the doubling of one's salary. Another study found having a friend whom you see on most days, compared to not having such a friend, had the same impact on well-being as making an extra $100,000 a year. Being married is also worth an extra $100,000, while being divorced is on par with having your salary slashed by $90,000. Just seeing your neighbor regularly is like making an extra $60,000. By far, the most valuable nonmonetary asset researchers examined was physical health, with "good" health compared to "not good" health equivalent to about a $400,000 salary bonus. That might seem crazy, but if you were not in good health, how much money would you be willing to give up to be in good health again? The reason I mention health is that social factors are also huge determinants of physical health. Thus social factors determine well-being directly and, because they bolster health, provide an additional indirect route to well-being.

The good news is that building more "social" into our lives is very cost-effective—getting coffee with a friend, talking to a neighbor, or volunteering won't make your wallet light and could significantly improve your life. The bad news is that as a society, we're blowing it. Over the last half-century, there has been a steady decline in nearly all things social apart from social media. People are significantly less likely to be married today than they were fifty years ago. We volunteer less, participate in fewer social groups, and entertain people in our homes less often than we used to.

To me the most troubling statistics focus on our friendships. In a survey given in 1985, people were asked to list their friends in response to the question "Over the last six months, who are the people with whom you discussed matters important to you?" The most common number of friends listed was three; 59 percent of respondents listed three or more friends fitting this description. The same survey was given again in 2004. This time the most common number of friends listed was *zero*. And only 37 percent of respondents listed three or more friends. Back in 1985, only 10 percent indicated that they had zero confidants. In 2004, this number skyrocketed to 25 percent. One out of every four of us is walking around with no one to share our lives with. Being social makes our lives better. Yet every indication is that we are getting less social, not more.

Why Are We Getting Less Social?

It shouldn't be surprising to us that being social is essential to our well-being. Everything we have learned about the social brain tells us that we are wired to make and keep social connections, that we feel pain when these connections are threatened, and that our identity, our sense of self, is intimately tied up with the groups we are a part of. As we've seen, our brains naturally gravitate to the social. Yet as a society we have been gravitating away from all things social. For thousands of years, we lived in small communities where we knew our neighbors and everyone around us because the communities were highly stable. Something has changed dramatically in the last century, something that is making us less happy than we used to be—less happy than we could be.

Unfortunately, my own life is pretty illustrative in explaining what has happened to many of us. I grew up in New Jersey, went to college there, and made a great group of friends. Then I moved to Massachusetts for graduate school and largely lost contact with my college buddies. Then I moved to California to become an assistant professor at UCLA. I lived in West Hollywood, where I made some good friends, but I lived far from campus. Once Naomi, my future wife, and I got serious, we moved much closer to campus, and I pretty much stopped seeing friends from West Hollywood. You've heard about how bad the traffic is in LA? Well, it makes you not want to drive eight miles on a Friday night. Plus, I was a new professor, working my tail off to get ahead like anyone who is new to their profession. Then Naomi and I got married and had a son, and I really wanted him to have a backyard to kick a ball around in or shoot hoops. So we bought a house, which led me to take on consulting work on top of my day job.

No need to feel sorry for me—I'm extremely lucky. I'm married to my best friend in the world and love my family, both immediate and extended. But when I look back over the choices I have made, apart for the brilliant one where I asked Naomi to marry me, I have made a series of choices that have moved me geographically and emotionally away from friends and have taken time away that could have been spent with loved ones. Without realizing it, I had moved from being a philosophy major, explicitly eschewing material pursuits, to an adult pursuing the "American dream." Somewhere along the line, the pursuit of happiness got confused with the pursuit of income and career advancement.

As in my own life, materialism in our culture has been growing over time, and this aspiration toward financial success for many of us has come at the cost of our social connections. We have limited time, and spending more time working means less time socializing. In 1965, only 45 percent of college freshmen listed being "very well-off financially" as a top life goal. At that point, "helping others" and "raising a family" scored higher. But by 1989, being well-off was at the top of the list, with 75 percent endorsing it. And this is sobering news because the more individuals endorse materialism as a positive life value, the less happy they are with their lives.

Back to School

Increasing the social connections in our lives is probably the single easiest way to enhance our well-being. But a growing addiction to materialistic values is taking us in the wrong direction, causing us to sacrifice time and energy from our social lives to the pursuit of financial success. I think it's safe to say that the government and corporate organizations that run our country have little interest in scaling back materialism—it grows the tax base and adds jobs, as more people are needed to make new things that people want to buy. After 9/11, President Bush's advice to the American public was to "go shopping." From the perspective of well-being, the government's interest in increased consumerism is largely a Ponzi scheme—it promises increased happiness but doesn't deliver. Regardless of our take on materialism, society should be deeply concerned about how we turn around our march toward social isolation. When we're socially connected, we are happier, healthier, and better citizens.

In the 1950s, the U.S. government took on numerous initiatives to build the physical infrastructure of the nation. Best known is the Federal-Aid Highway Act signed into law by President Eisenhower, which devoted over $400 billion (in current dollars) to the creation of more than 40,000 miles of interstate highways. This investment has been repaid many times over in the form of new economic activity. When the Great Recession hit in 2008, lawmakers quickly drew up plans to rebuild the nation's now crumbling infrastructure. Roads and bridges are in dangerous disrepair. Our railroad system is far behind the rail lines of many other modern nations. Rebuilding the infrastructure would create jobs and ultimately spur new economic activity.

I would argue that what we need is a new stimulus to rebuild the social structure of our society, as well. To be fair, the government does make major investments in social programs. But these are safety nets, rather than programs to increase social connection. Programs like Social Security and Medicaid provide some measure of financial and physical security to those who are less able to fend for themselves. But these are not investments in enhancing the social lives of all citizens. Yet such investments would likely be repaid through higher productivity, better health, and lower crime. Unfortunately, because social connections are less concrete than a new highway, it may be hard for people to rally behind them. Still, we know now that our brains are wired for social integration and that this wiring permeates virtually every aspect of our lives. Imagine if the president created a Council of Social Advisors to parallel his Council of Economic Advisors. Bill Gates has been convincing the world's billionaires to donate much of their wealth to supporting worthy causes such as ending polio. What if they invested a little bit in social well-being?

Many of us can remember back to when we lived in a dormitory, freshman year of college. Think about the extraordinary feat of social connection that happens on dorm floors each year. Incoming students arrive at college in a socially vulnerable state, often with no pre-existing friends at the school. Dorm floors are ground zero for early social connecting in college. Many of the people on each floor will make close friends with one another, and some of these friendships will last a lifetime. Apart from the military, I can't think of many other institutions in our lives that are as conducive to the creation of social bonds.

Approximately a third of all Americans live in apartments, which in most cases are physically similar to college dorms. Yet living in an apartment feels nothing like living in a dorm. So what is it that colleges get right in how they set up their on-campus communities? It certainly isn't the food or the luxurious size of the dorm rooms. First, I think, they get the physical space right from a social perspective. When I was an undergrad at Rutgers, each dorm floor devoted about 20 percent of its space to areas for social gathering. Dorms have couches with cable TV, and these days, some have videogame systems as well. I have lived in several apartment buildings in my life, and I have never seen one with any meaningful common space devoted to floor-specific socializing. Some have a sizable lobby, but they are not set up for informal socializing. Open spaces on each floor work in part because people can hear what's going on on the floor or in the building, and they can casually walk by and check it out. Of course, the space would need certain amenities to attract people to it. People might claim they head there for the large-screen TV or free Wi-Fi, but they would probably stay for the socializing.

The reason colleges get this right and apartment buildings don't is that they have different motives. Colleges are concerned with having a vibrant community; apartment builders are concerned primarily with profit and the costs per square foot. But as a society, shouldn't we too be concerned with having vibrant communities? Given that 100 million people live in apartments in the United States, structural solutions to improving their social lives would seem like a good investment for all of us. Couldn't we, for example, offer tax breaks to those who agree to build one less apartment per floor and leave the space open for socializing? In other words, doesn't it make sense to take what we know about the value of social connection and use this to guide how a portion of our taxes is used?

Colleges have other tricks up their sleeves to encourage socializing. For many schools, students fill out profiles about their likes and dislikes, which are then used to pair up roommates. While this doesn't apply to apartments directly, this technique could be used to match incoming tenants with someone else in the building who has some similar preferences or is at a similar lifestage (that is, raising a newborn, just retired, and so on). Last, but not least, each college dorm floor usually has an older student living rent free in return for overseeing the floor and creating a series of social activities. These begin with "get to know each other" events at the beginning of each year and then move on to things like movie night, poker games, and board games night. Students wanted to socialize, but didn't always know how to get the ball rolling, and that's where the dorm advisors come in.

Throughout our childhoods and young adulthoods, our social lives are curated by others. Couldn't we find a way to replicate that in our adult communities as well? Why don't we have someone on each apartment floor designated to create social activities? In a sizable apartment building, it wouldn't be hard to raise $1,000 per floor per month, taking a sliver of the rent from each apartment. This money could be split between funds to support the activities and funds that would be used as payment to the social organizer on the floor.

In the neighborhood where I live, we have an organization of property owners that typically serves two functions. The first is political, fighting for various things like more police cars on our streets. But the second is informational. There is an Internet listserv that allows residents to ask questions like "Does anyone want to buy my tickets to the Lakers game next week?" or "Anyone know a good plumber?" Why not extend that concept further? For example, close off streets one evening each weekend so that communities can use the streets themselves to set up a variety of social events.

Snacks and Surrogates

Knowing what we know about the brain's wiring for social connection and how social connection relates to well-being, shouldn't we look to change our schedules to work less and socialize more? Research suggests that when people think about money, they become motivated to work more and socialize less. But when people are prompted to think about time, the reverse happens; people become motivated to work less and socialize more.

People also find ways to extract some of the benefits of socializing even when there is no one around to be social with. Social psychologists Wendi Gardner and Cindy Pickett suggest that people can extract some of the benefits of socializing through *social snacking*. Just thinking about or writing about a loved one can provide some of the benefits of face-to-face social relationships. Looking at a picture of a loved one can offer some of the benefits of traditional social connections.

Social support and social connection can buffer us against the stress of difficult moments in our lives. In one study that Naomi Eisenberger and I ran, when we delivered a painful stimulus to women, they reported the pain to be less painful when they were holding their boyfriend's hand. Surprisingly, when the girlfriend was merely shown a picture of her boyfriend, the pain was still reduced. In fact, the picture was twice as effective in reducing the women's pain as actual handholding.

In other words, a picture of a loved one is a strong enough social reward to help overcome some kinds of distress. Inspired by our work, Nikon recently partnered with the Red Cross in Germany to bring digital picture frames to people in hospitals so that they would be able to see pictures of their loved ones during their hospital stay.

Television is the number one leisure activity in the United States and Europe, consuming more than half of our free time. We generally think of television as a way to relax, tune out, and escape from our troubles for a bit each day. While this is true, there is increasing evidence that we are more motivated to tune in to our favorite shows and characters when we are feeling lonely or have a greater need for social connection. Television watching does satisfy these social needs to some extent, at least in the short run. Unfortunately, it is also likely to "crowd out" other activities that produce more sustainable social contributions to our social well-being. The more television we watch, the less likely we are to volunteer our time or to spend time with people in our social networks. In other words, the more time we make for *Friends*, the less time we have for friends in real life.

Over the last two decades, the Internet has increasingly been challenging television for our leisure time. As with television, people have turned to the Internet to fulfill their social needs. Unlike television, which is a passive activity, the Internet offers endless opportunities to actively connect with other people. Although people avail themselves of this online social mecca, there have been major questions about its utility. Does more time spent online lead to better well-being the way offline social connection does? And how does time spent online affect our social connections in the "real world"?

In 1998, Robert Kraut published the first seminal study examining these questions. What he found was disturbing. Individuals who used the Internet more decreased communication with their families, had shrinking social networks, and reported increased depression and loneliness. A series of other papers came out soon after confirming these and other negative consequences of Internet usage. But a few years later, a strange thing happened—all the new data was starting to show positive effects of Internet usage on social connection and well-being. Why the change? In a word, Facebook. In six words, people started using the Internet differently.

In the 1990s, the social usage of the Internet centered around topic-specific chat rooms. People with common interests would enter the same chat room to discuss mutual interests. People were going online to find new people who cared about what they cared about, and sometimes, after connecting online, these people might connect offline as well. In most cases, though, people in these chat rooms did not connect in real life.

Facebook, created in 2004, was originally designed to facilitate an existing community—undergraduates living on the same campus. It was a complement to real-world social connections rather than a replacement. Although Facebook has grown beyond anyone's wildest dreams, the original functionality is there. People consistently report being more motivated to use Facebook to connect with people from their offline lives than to meet new people. Because Facebook use is more of an extension of real-world connections, it has been associated with enhancing offline social networks and general well-being. It's also particularly useful for maintaining social bonds over long distances. Even though I live thousands of miles from the places where I went to college and graduate school, it's easier to keep connected with these friends than it would have been decades ago.

Social snacks and surrogates are a testament to the power of social motivations. People need to connect and have discovered all sorts of ways to satisfy this urge. Some work better than others, and still others remain to be invented (holodeck, anyone?). We should make use of these ways of connecting or savoring connections because they make us happier and healthier.

CHAPTER 11

The Business of Social Brains

work in a massive organization (the University of California), and I manage a team of graduate students, postdoctoral fellows, undergraduate research assistants, and staff. Working in this kind of organization is both a blessing and a curse. It's a curse because of the endless bureaucratic red tape we deal with, being a public institution. The primary blessing, beyond the sheer brilliance of the people I get to be around, is that we think of ourselves as a family. I spend a lot of time thinking about each of my students and whether they are on the path to future success. It's similar to how I think about my own son, Ian, and whether he is developing in ways that will allow him to be successful in his own adult life. Indeed, I refer to Dan Gilbert, my PhD advisor, as my academic father, and my students are his academic grandchildren. Treating my lab like a family has had enormous benefits for how the lab runs. Unfortunately, most businesses, particularly larger ones, do not operate this way most of the time.

Working in groups and organizations, is a fact of life for most adults. It's the engine of economic growth for our society, the source of our incomes, and often it's the place where we spend most of our waking hours. Yet most organizations don't get "social" right. They don't feel like families and they don't feel like a positive part of one's social life. Given what we know now about the social brain, creating the right social environment in our places of work should be a top priority for anyone who wants the best out of themselves and those around them.

Don't Forget Your SCARF

If you run a company or department and want employees to show up on time, work harder, and stay with your company longer, there is a tried-and-true solution: pay people more money. As economist Colin Camerer wrote, "Economists presume that [people] do not work for free and work harder, more persistently, and more effectively, if they earn more money for better performance." Of course we have already seen that making more money doesn't actually make people much happier. But people *believe* money will make them happier. So offering more money for greater productivity ought to motivate people. This has increasingly led to an incentive system where higher pay or bonuses are available, contingent on performance. Pay for performance intuitively seems like a great idea. Indeed, at first blush it might seem like the *only* idea when it comes to increasing productivity. But in fact, higher pay often turns out to be a poor investment. There are certainly some contexts in which financial incentives increase performance, but there are others in which money actually undermines performance. Generally speaking, pay for performance usually produces small or no performance improvements.

Despite this disconnect, pay for performance is the dominant model used in the business world to increase productivity. When all you have is a hammer, everything looks like a nail. When money and the physical comforts it can buy are the only things you think motivate people, it will be offered as the solution to every workplace problem.

But we know better. We know the brain is wired to care about social pains and pleasures in much the same way it cares about physical pains and pleasures. These are core features of the brain.

David Rock, who runs the Neuroleadership Institute, has spent the last decade persuading businesses that "social" matters in the workplace—that if it is incorporated into workplace practices, it can produce a far better work environment that enhances employee engagement and productivity.

Rock has developed the SCARF model. The acronym stands for *status, certainty, autonomy, relatedness,* and *fairness.* They are the nonmonetary drivers of behavior, "the primary colors of intrinsic motivation" according to Rock.

I think it's a great place to begin because these are all part of our basic motivational machinery. Each of these can lead to what will look like irrational behavior to observers who think pain and pleasure are the only rational sources of motivation. Autonomy and certainty aren't really a social part of the story, and they have been described well elsewhere (*Drive*, by Daniel Pink). However, status, relatedness (or what I call *connection*), and fairness are phenomena that can trigger social pain and pleasure responses in the brain.

Imagine the face of a CEO of a Fortune 500 company being pitched the idea that in order to motivate his employees he should focus less on financial incentives and more on status, relatedness, and fairness. His expression might reveal contempt or confusion as to why you would make such a ridiculous claim. He might go on to ask something like this: "How can social connection produce the same the bang for your buck as an actual buck?" Here is the answer: We evolved over millions of years to become a deeply social species. This means that a variety of motivational mechanisms predispose us to respond positively to signs that we are accepted by the group (*connecting*), and it also means that we are motivated to work hard for groups we identify with (*harmonizing*).

In trying to persuade our CEO, let's start with the easy non-monetary motivator: status. This one is easy because our CEO will say, "Sure, I already know about status." But his theory about status will mostly likely be the wrong one. He assumes that people seek out status because it's a sign that more money is likely to follow.

Status is a means to a means to an end—physical comfort. But evidence suggests that status is also an end in itself. We desire status because it signifies that others value us, that we have a place of importance in the group and therefore are connected to the group.

A recent study demonstrated how much people crave status and recognition in the workplace, even when there is no material payoff from it. The title of the paper says it all: "Paying $30,000 for a gold star." Economist Ian Larkin took advantage of a peculiar situation at an enterprise software vendor to separate recognition from the increased money that typically accompanies it. Some companies acknowledge individuals who end the year in the top 10 percent of sales with membership in a "President's Club," which typically has few tangible benefits. In the case of the company Larkin was studying, the names of the winners were shared in an e-mail from the CEO to all employees, a gold star was put on their business cards and stationery, and all the winners took a three-day trip together to an island resort (valued at $2,000).

Salespeople close to the 10 percent level heading into the last months of the year faced a dilemma. If they completed their deals in the current year, they would dramatically increase their chances of getting into the President's Club. But this would cost them a lot of money in commissions. The company has a commissions accelerator program: the higher one's sales in a quarter, the higher the commission rate. For instance, a sale made in a poor quarter might yield a 2 percent commission, but in a strong sales quarter, the same sale might yield a 24 percent commission. A person who has a number of sales lined up toward the end of a weak quarter would be wise to delay those sales until the next quarter, when, combined with the normal sales from that quarter, they would end up yielding higher commissions. Indeed, this is exactly what people who aren't close to getting into the President's Club do. But salespeople close to getting the "gold star" face a quandary: whether to finish up the sales in the current quarter and get the recognition, or move the sales into the next quarter and make more money.

Of the people facing this dilemma, 68 percent chose to take the immediate sales in order to ensure that they would get into the President's Club. By doing this, they gave up about $27,000 in future commissions on average (far more than the value of the three-day vacation). The affected individuals earned about $150,000 in yearly salary and commissions. In other words, they were willing to trade 20 percent of their salary for the privilege of being recognized as a high-status salesperson. Larkin tracked whether these individuals benefited in terms of future sales or promotions. They didn't. All they got was the recognition itself. When confronted with the results, at least some of the employees felt they had made the right decision. One said, "I paid $20,000 for that Gold Star. And it was worth it." When we hear this statement, we think that the employee must not really mean it—it's just rationalizing their irrational behavior. But if we understand that status might act like an ongoing trigger to the brain's reward system, then it starts to make more sense.

Does our hypothetical CEO have a Gold Star program in his company? He should, after hearing about Larkin's research. Recognition is a free renewable resource. I can't imagine many CEOs who would prefer giving out large bonuses instead of patting a deserving employee on the back. In Larkin's study, the $27,000 that employees gave up didn't vanish into thin air. It went straight into the company's bottom line—as profit.

On to connection. If you take a moment to think it through, the benefits of social connection (or relatedness) in terms of productivity are self-evident. When you work on a project in an organization, most of the time you will not be able to complete the project on your own from start to finish. Either you will be assigned to work on it with a team, or you will need to reach out to others to help you with certain aspects of the assignment. Let's say you need someone to put together specialized analyses to include in your report and you can't move forward until you receive them. Who is going to get them to you more quickly, a friend or a stranger? And if you were the person being asked to provide the analyses, taking time away from your own work, whom would you be more motivated to help?

In my own lab, I strive to admit graduate students who are socially skilled, in addition to being smart and motivated. The graduate students in my lab all have their distinct areas of expertise, so they all need to be comfortable learning from others and teaching others—helping and being helped. Being smart and motivated, without being able to connect with others in the lab just won't cut it. I've had a couple of students in the lab over the years who never really integrated socially with the rest of the team, and they often struggled. They could leverage their own intelligence and hard work, but they were less able to access the intelligence and expertise sitting in the next office over. From this perspective, social connection is a resource in the same way that intelligence or the Internet are resources. They facilitate getting done what needs doing.

Economists have long studied human capital as a driver of productivity in organizations. *Human capital* is the amount of intelligence, experience, and education a person has. Not surprisingly, companies with more human capital tend to do better. However, most studies of human capital ignore the concept of *social capital*, the social connections and social networks within an organization. Does human capital lead to productivity all on its own, or does social capital play a role in catalyzing human output into optimal performance?

Economist Arent Greve studied three Italian consulting companies to find out. He measured the human capital and social capital of the employees at these companies and then related that finding to how many projects each person completed in a year as a measure of productivity. Bottom line, in two of the companies, social capital accounted for all the benefits in productivity. In the third company, human capital did have an effect, but this effect was augmented to the degree to which a person also had strong social capital. The assumption that productivity is about smart people working hard on their own has been masking the fact that individual intelligence may only be optimized when it is enhanced through social connections to others in the group. Social connections are essentially the original Internet, connecting different pockets of intelligence to make each pocket more than it would otherwise be by itself. These social connections turn out to be even more important for small companies and start-ups that specialize in innovation.

Even a social factor as innocuous as fairness in the workplace can significantly affect job performance, absenteeism, turnover rate, and organizational citizenship. The extent to which employees perceive decisions to be fair in their place of work can account for 20 percent of the differences in their productivity. I'm not aware of any analysis suggesting that financial incentives can have nearly the same impact. Fairness might seem like a squishy motivator, but recall that fairness activates the same reward circuitry in the brain as winning money.

Status, connection, and fairness all have demonstrable effects on the bottom lines of organizations. Yet few take these issues seriously. Enhancing these factors are low-cost, efficient strategies for improving workplace outcomes. Whether employees realize it or not, they have been wired to be motivated by being accepted and valued by the groups they are socially connected with.

Care to Succeed?

The SCARF model is a great way to keep track of the things that motivate us other than money and physical comfort. That said, there is one more particularly counterintuitive social element to add to the ideal work environment: opportunities to care for others. When we discussed social rewards in Chapter 4, we used the parent-child relationship as our jumping-off point. We discussed how there are two sides to this equation and thus two kinds of social rewards that activate the brain's reward circuitry. As children, we are built to be sensitive to cues that we are liked, loved, and cared for. As we grow older, being respected and valued increasingly matter as well. However, as parents and as adults more generally, we are reinforced by the actions we take to care for others.

This is probably the hardest social factor for our CEO to wrap his head around. It is a bizarre-sounding incentive.

Adam Grant, a professor at the University of Pennsylvania, has done fascinating research, described in his book *Give and Take*, showing that the chance to help others motivates people to work harder in the workplace. He has taken two different, but complementary, approaches. In the first, he focused on meaningfulness in the workplace. Ever since Maslow's hierarchy of needs, people have suggested that we are more motivated to do things that are personally meaningful to us. Grant's big insight was that for most people in most lines of work, doing something meaningful means helping others. It's hard to find meaning in what we do if at some level it doesn't help someone else or make someone happier.

Of course, we don't all find meaning in what we do. And because of mass production, it has become harder and harder to find meaning in our work; so many of us add only a tiny part or contribution to some overall process or output. With the rise of the Internet, we are even less likely to come face-to-face with the people who eventually benefit from our work.

Grant's studies introduced interventions that made employees more aware of how their work helped others. In his first study, he focused on people working at a university making calls to alumni to try to raise money for undergraduate scholarships. This is a hard job because people usually don't want to be bothered for money over the phone; many are saying under their breath, "I gave you four years of tuition! Isn't that enough?" Callers must be focused on how to keep prospective donors on the line long enough to make their pitch; they don't have a lot of time to think about the ultimate beneficiaries of their efforts. Grant gave some of these callers a surprise visit with a past scholarship recipient, someone who directly benefitted from the work the callers do. The visits weren't long—just five minutes. At the end of the five minutes, the manager came into the room and said, "Remember this when you're on the phone—this is someone you're supporting."

You probably think this made the callers feel pretty good for a little while, maybe even the rest of the day, but that it likely had no lasting effect. But you would be wrong. To test the effects, Grant got access to the callers' performance data from a week-long period before the meeting with the scholarship recipient and then again from a week-long period *a month after* the meeting. The work performance of callers who hadn't met a scholarship recipient was about the same in both time periods. They spent roughly the same number of minutes on the phone trying to get donations and brought in roughly the same amount in donations. The work performance of those who met the undergraduate was radically different in the two periods. They increased their time on the phone 142 percent from the week before the meeting to the week-long period a month later. And this led to much greater success. Donations for these callers went up 171 percent across the two time periods!

Have you ever heard of such a brief intervention having such profound consequences for work performance? All Grant had to do was remind people of how their work was helping others. The job was the same, but clearly the callers' psychological mind-set changed, and it remained changed for a long time. Remember that the second measurement came a full month after the meeting with the scholarship recipient.

In follow-up studies, Grant replaced the face-to-face meetings with letters that described how the employees' work benefitted the scholarship recipients. He compared changes in the performance of these callers relative to the performance of callers who received letters describing how their work benefitted them personally. Those who read the letters that described how their work benefitted them personally did not change their performance; but those reading about how their work helped others showed dramatic gains. The number of donation pledges obtained increased 153 percent, and the total value of their pledges increased 143 percent. All this just from a letter.

Grant's second approach to caring and workplace success involved a very different kind of caring—giving support to fellow employees. Many companies now have employee support programs designed to support employees in nontraditional ways. These programs provide child or elder care and, for employees in need, direct financial assistance from the company. In some companies, including Southwest Airlines and Domino's Pizza, employees are also given the opportunity to *donate* to these programs in order to help their fellow employees in times of need. Grant examined one of these programs at a major retail company. Although he did not measure job performance directly, he did examine employee engagement, which is a good proxy for job performance. Those who helped out with the employee support programs, either donating money directly or volunteering time to help raise money for the program, reported feeling more engaged and committed to the company.

The axiom of self-interest suggests that a person who helps someone else out always does so with the expectation of getting something of equal or greater value in return. When an employee contributes to the employee support program, he is helping both the company and another employee. In return, the helpful employee might then feel entitled to slack off a bit ("the company owes me") and perhaps even expect the employee who received help to pick up the slack ("he owes me"). Instead, the employee who donates his time or money finds his engagement with the company goes up, meaning greater productivity, reduced absenteeism, and lower turnover. Benjamin Franklin understood this long ago, writing, "He that has once done you a kindness will be more ready to do you another than he whom you yourself have obliged."

How are we to make sense of this? There are several reasons why an employee who donates to an employee support program might become more productive. The first reason is self-perception. When we see ourselves doing something, we tend to infer that it reflects on who we are in general (especially if it's a good thing). This makes us more likely to do additional things consistent with that view in the future. Donating to an employee support program makes employees more likely to see themselves as good citizens of the company, and working hard is another behavior consistent with that self-view. A second explanation is that helping others makes us feel good: it activates the brain's reward circuitry, making it more likely that we will feel positively toward the organization that gave us the opportunity to experience those good feelings.

A third explanation is that we are motivated to help and we value others who demonstrate that they too like to help others. We like to see others showing they care. Employee support programs demonstrate that a company cares about its employees. People who donate to these programs have probably spent more time than others thinking about the fact that the company is a caring organization. As one of Grant's participants said, "I do feel very attached to the company. . . . I always feel proud that the company supports the employee support program." Just as in any family, people who are strongly attached will work harder to support the family and help it thrive. This is no different in the organizational context.

Attachment, the kind Bowlby and Harlow talked about, matters in the workplace. People often talk as if their company, job, or workplace is solely about getting a paycheck and helping the company increase profits. This conversation is predicated on the norm of self-interest—the belief that material self-interest is the only thing that motivates people, individually and corporately. We have been bombarded with this idea for so long that it's the only conversation we know how to have about the workplace. But it is the wrong conversation to be having because it misses so much of what actually makes us us.

Material self-interest is pervasive, and most people can't afford to work for free. However, most of us spend a quarter of our adult lives (40 hours in a 168-hour week) working, which means all of the social motivations wired into our brains will be expressed at work too. Knowing we are in an organization that cares for us, for other employees, and for the community creates attachments that are surprisingly effective at keeping us engaged and motivated. Few of us know this about ourselves, but that doesn't make it any less true.

Building a Better Boss

A recent poll asked employees which they would prefer, a raise or a better boss. Two-thirds (65 percent) answered that they would prefer a better boss to a higher salary. Some managers might feel that being disliked by one's employees is a necessary cost of squeezing maximum productivity out of people, but Gallup recently estimated that these poor manager-employee relations are costing the U.S. economy $360 billion a year in lost productivity. Employees who are unhappy at work slack off in hard-to-detect ways and are less likely to generate and share new ideas.

No one told me that I would someday be a boss, running my own small business, but that's exactly what running a scientific laboratory is like. Every year, thousands of psychology graduate students get their PhD, but only a small fraction of those are chosen to become bosses (that is, professors) who will run their own lab. The odd thing is that the skill set that is essential for *becoming* the boss (that is, publishing high-quality research as a graduate student) has little to do with the skill set necessary for *being* the boss.

Graduate students study different psychological phenomena and hope that a few of their studies will work well enough and be interesting enough that a university will want to hire them to do more of that work. Beyond luck, most of what helps a graduate student get ahead is being really smart, developing technical and content expertise, and working really hard—having the intellect and focus necessary to give luck a chance to do its thing.

Intellect, expertise, and focus are still essential parts of being a professor, but a huge part of my job has become managing the folks in my lab. I have had to solve countless social and motivational problems over the years. Doing this requires appreciating the complex social dynamics going on between members of my lab and what being in the lab represents for them in terms of their current and future identities. Getting my work done depends on them getting their work done. Getting their work done depends on my understanding their needs, what motivates them, and how to create the best work environment. No one ever discussed this with me in graduate school. There were no classes on how to do this well. When I interviewed for the job, being able to manage the social dynamics of a lab never came up. And while I have muddled through, I wouldn't necessarily give myself high marks for the social side of being a leader.

For better or worse, I am hardly alone. People are moved into managerial roles because they were the most skilled, intelligent, or productive team member in a nonmanagerial position. If you have a dozen engineers working in a group and the manager leaves, creating a leadership vacuum, it is natural for the organization to promote the most successful of the dozen to become the new manager.

Catalyzing Leadership

I've given you my own anecdotal experience, but are social motivation and social skills actually important for a leader's success? If they are, then why don't we see more bosses being selected and promoted for their social competence?

Answering the first question is straightforward. The social ability of leaders can have huge consequences for the success of their teams. John Zenger, a leadership expert, has asked thousands of employees to score the leadership effectiveness of their boss. He found that when he divided the bosses into "great" (top 20 percent), "good" (middle 60 percent), and "bad" (bottom 20 percent), the result was highly predictive of various outcomes including profit, employee satisfaction, turnover, and customer satisfaction. He then described five leadership competencies that he hypothesized would be associated with being a better leader: personal competence (intelligence, problem solving, expert knowledge, and training), focus on results (being driven to move tasks forward and complete them), character (integrity and authenticity), leading organizational change, and finally, interpersonal skills. His analysis then focused on pairing different competencies together to improve overall leadership. Zenger found that combining interpersonal skills with other competencies allowed leaders to maximize their effectiveness.

Zenger found that if employees rated a manager as very high on "focus on results" (that is, one's ability to get things done effectively), there was still only a small (14 percent) chance that the manager would be rated among the top 10 percent of leaders overall. However, if in addition to "focus on results," employees also rated the manager's ability to "build relationships" very highly, then the likelihood of that person's being rated as a great leader overall skyrocketed to 72 percent.

Essentially, social skills improve the value of the other competencies because they allow leaders to manage the social and emotional responses of their employees. When employees are performing some aspect of their job incorrectly, there is a fine line between correcting them in a way that is supportive and correcting them in a manner that makes them feel rejected, undermining their willingness to internalize the feedback and to work hard in the future. Social skills allow the boss to walk that tightrope without falling off.

Sometimes social skills are more important than personal competence. In a laboratory study, three-person teams were brought together to perform a complex task for which it was natural for someone on the team to emerge as a leader. At the end of the task, each team member rated the others on the extent to which they emerged as an effective leader. A team member's intelligence and social skills were both associated with being singled out as an effective leader. However, social skills were considered nearly twice as important.

If social skills exert such a strong influence on the success of leaders, they should be a major criterion in the hiring and promotion of managers and executive leaders within a company. Unfortunately, they are all too often overlooked. David Rock, who works with dozens of Fortune 500 companies, sees this all the time. "One of the most common concerns I hear from organizations every week is that the more technical their people are, the worse their social skills seem to be, and that this can really become a problem when they become a manager or leader." In a recent survey conducted by the Management Research Group and the Neuroleadership Institute, the competencies of thousands of employees were examined. Although more than 50 percent were rated by their bosses and peers as having a high degree of "goal focus," less than 1 percent were rated as high on both goal focus and interpersonal skills. We know from Zenger's analyses that putting these two competencies together is essential to leadership success, but it is clear that businesses are neither identifying individuals with both nor cultivating leadership through their culture or training programs.

The Neural Seesaw

Why aren't leaders always selected with social ability in mind? One of the reasons for this is that our mental representation of what a leader looks like is at odds with what actually makes for a successful leader. Robert Lord has studied perceptions of leaders for decades. In one review of more than two dozen studies, he examined the characteristics that people associated with leaders, in order to identify which traits came to mind most frequently. He found that "intelligence," "dominance," and "masculinity" were consistently rated highly; social skills didn't make the cut. People think of leaders as smart and forceful, rather than as socially skilled. This perception no doubt influences hiring decisions.

In addition to issues of perception, perhaps there is something about the fundamental relationships between analytical and social intelligence that makes it more difficult to identify leaders who show strengths in both. One study examined this possibility by looking at the relationships between intelligence, empathy, and leadership. Intelligence and empathy were each associated with leadership; however, intelligence and empathy were negatively correlated with one another.

We have already seen this trade-off between social and nonsocial thinking in the moment-to-moment dynamics of the brain. Recall the mentalizing network that allows us to think about what is going on in the mind of others (see Figure 2.1). There is also a separate network for abstract reasoning about nonsocial phenomena that is associated with general intelligence (see Figure 2.3). One of the defining features of these two networks is their relationship with each other. When we are left to our own devices to think as we please, these two networks act *like two ends of a seesaw*; as either side increases (goes up) in activity, the other side decreases (goes down).

This relationship between thinking socially and thinking non-socially may make it hard to do both at the same time. In many cases, mental processes facilitate one another rather than competing. For instance, seeing and hearing complement one another. Seeing someone's lips move as they speak helps our auditory processes unpack what we hear that person saying. Though there have been studies showing the social and nonsocial reasoning systems operating in a complementary fashion, it is far more common to see them at odds with each other.

There are two ways to think about this antagonism between social and nonsocial intelligence as it relates to leadership. First, some people might just have an enduring predisposition to activate the network for nonsocial reasoning, deactivating the social network as an accidental by-product. This could be a result of genetics or the result of a lifetime of practice, living in a society that values abstract thinking over social thinking.

Alternatively, other people might prioritize nonsocial thinking because of how they think about their job. To the extent that someone frames a leadership task primarily in nonsocial terms, they are more likely to suppress the social mind, rendering them less sensitive to the social events around them and less likely to consider the social implications of their own behaviors and those of their employees. Often when a team member says that she is having trouble making progress on a task, the subtext may be that she is having difficulty working well with one or more other people on the team. A leader who is socially attuned may realize the group dynamics need work. A leader who isn't may focus on whether the employee needs more personal training in order to be able to complete the task—a poor solution to the actual problem.

There's good news and less good news when it comes to dealing with the brain's seesaw between thinking socially and thinking nonsocially. For those who frame aspects of their work as fundamentally nonsocial, a shift in how the task is understood may lead to more balance. The most effective leaders are able to bounce back and forth between these mental modes. That's the good news. The less good news is that if a person is biologically disposed to favor the nonsocial network, a simple reframing of the job is unlikely to do the trick. For someone who has spent a lifetime overlooking the social aspects of the workplace environment, becoming fluent in social understanding might be akin to learning a second language in adulthood. It can be done, but it takes a lot more effort than it would have taken in childhood.

The best bosses understand and care about the social motivations of all the members of the team. Bosses have to foster better *connections* between themselves and their team, among team members, and between the team and other outside groups and individuals critical to success. Better communication will reduce the *mindreading* burden on everyone on the team, and it will allow social issues to be nipped in the bud, rather than festering from one project to the next. Efforts to make the group actually feel like a group will be rewarded, as team members start to better identify with the team. This will facilitate the kind of *harmonizing* that will promote individuals thinking about how they can best serve the team, rather than themselves. As social animals, we are wired to do this, as long as we really identify with the team. Creating this identification, this attachment to the group, is an essential component of successful leadership.

All of this is really just the start of a conversation about how the social brain influences the workplace, from work spaces to organizational structure. But it is a conversation that hasn't been had enough. And done well, it can transform an organization.

CHAPTER 12

Educating the Social Brain

In the United States, we spend more on public education (kindergarten through twelfth grade) than nearly any other country (more than $800 billion per year). And yet international comparisons suggest that our students are lagging behind most industrialized nations in math, science, and reading. Out of 34 comparison countries, U.S. students rank twenty-fifth in math, seventeenth in science, and fourteenth in reading. This means that as a country, we are getting a lousy return on our investment in education.

My belief is that junior high holds the key to our educational woes. Junior high is made up of seventh and eighth graders who are between the ages of twelve and fourteen. There is a drop on several key educational indicators that occurs between fourth and eighth grade, and if we can stem the tide of disinterest and disengagement that occurs during these years, I think the societal payoff would be immeasurable. There are few problems that, if solved, would have more widespread benefits than keeping our children interested and excited about their own education.

The primary pathway to solving this problem over the last decade has been the accountability approach embedded in the No Child Left Behind (NCLB) Act passed by Congress in 2001. This plan focused on annual testing of students and the creation of report cards to assess each school (thus, if the children fail, so does their school). There are various criticisms of NCLB, and the general consensus is that while it has increased performance on the specific tests associated with NCLB, it has not increased real learning or improved our international standings. For my purposes, it represents a strong contrast to approaches that incorporate what we have learned about the social brain. I would like to consider what our knowledge of the social brain tells us about how to improve education, particularly in junior high.

The Need to Belong

I began seventh grade as the prototypical "new kid," having just moved from one part of New Jersey to another. On the first day of class, I was lucky to make friends with a kid who liked the same sports I did, played videogames, and was really smart. In fact, he was so smart that I found it hard to believe he had done poorly on the placement tests we took the following week. When I asked him about it, he told me he had tanked the tests on purpose so that other kids wouldn't know he was smart and tease him. Although the geographical distance between my sixth and seventh grades was not that great, the universe seemed to have turned on its head such that it was suddenly very uncool to be smart and try hard. My friend was more concerned with being liked than with doing well. Thankfully, my friend didn't tell me any of this *before* the placement tests.

There are myriad reasons why academic performance and interest drop in junior high, but one nonobvious reason may be that the need to belong, our most basic social motivation, is not being met. Changing schools, from elementary to junior high, right around the same time that we reach puberty creates an uncertain and unstable social environment. This switch also brings with it a change from having a single teacher for most of the day, a teacher who knows each student well, to the high school model with different teachers for each subject.

Do junior high students feel like they don't belong? According to work by my UCLA colleague Jaana Juvonen, junior high students in the United States feel strongly that they don't belong. Juvonen and her team analyzed data from more than 32,000 junior high students across a dozen countries. On multiple measures, U.S. students reported feeling less socially connected to their schools, teachers, and peers than students in most other countries in the survey. Our students rated the overall school climate in junior high worse than did students in any of the other countries, and our ratings were twice as negative as the country ranking second to last.

The question of whether we should be investing our limited educational dollars in making junior high students feel more socially connected to a community depends on our goals for school and our beliefs about the role of social connection in obtaining those goals. Even tough-nosed cynics will likely agree that if we spend time and money creating a greater sense of community, our students will be happier. Without pausing to breathe, they might rush to add that student happiness is not the main goal of our schools. Although I care deeply about my own child's happiness and well-being, I agree that a school's main mission is to maximize learning and the capacity for self-guided learning later in life. The big question, then, is whether creating a sense of belonging in students is merely an end in itself or whether it can improve learning and educational outcomes.

Being Bullied

There is no bigger threat to an adolescent's sense of being accepted than being bullied while others stand by doing nothing. Observers' inaction is taken as a tacit endorsement and thus can be experienced as a broad rejection by peers. As might be expected, bullying in school is associated with negative changes in self-esteem, depression, and anxiety. Young adolescents experiencing more bullying or rejection at the hands of their peers also show subsequent decreases in their GPA and school attendance.

Given that as many as 40 percent of adolescents report being the victims of some kind of bullying, this may already be producing broad-based reductions in student achievement. Dewey Cornell examined how schools as a whole were doing on the various NCLB tests as a function of the prevalence of bullying in each school. Schools with higher bullying rates scored significantly lower on tests of algebra, geometry, earth science, biology, and world history.

Why would bullying, which typically takes place outside the classroom, affect performance in the classroom? Recall that social pain activates the same neural circuitry as physical pain. It is well established that chronic physical pain is associated with cognitive impairments like diminished working memory. The entire purpose of pain is to draw attention to itself so that corrective or recuperative actions can be taken. Thus, a person in pain is likely to be fixated on that pain, whether it's physical or social, and this focus would leave fewer cognitive and attentional resources free to focus on the lesson of the day.

Psychologist Roy Baumeister examined the hypothesis that social pain leads to decrements in intellectual performance. His team experimentally manipulated whether individuals were made to feel socially rejected or not. Then in different experiments, the participants took either an IQ test or a GRE-style test. In both cases the outcome was unequivocal. Social pain led to dramatic reductions in test performance. On the IQ test, the average score was 82 percent correct, but the scores of those who had been made to feel rejected fell to 69 percent. More dramatically, on the GRE-style test, the rejected participants barely got half as many right as nonrejected individuals (39 versus 68 percent). That is a stunning difference. And all Baumeister had to do to produce it was tell participants that in the distant future they were more likely to be alone than other people. Imagine the effect of being the victim of real bullying, particularly when no one stands up to take your side. This must be a profound distraction and a major strain on classroom learning.

Getting Connected

What about the flipside of the coin? Does feeling more socially connected increase academic performance? Do grades shoot up when we feel liked and respected? Just as our intuitive theories tell us that the negative consequences of social pains are stronger than the positive consequences of social connection, it has been more difficult for researchers to establish that enhanced feelings of belonging increase academic achievement. A number of studies have now shown a modest impact on GPA of being accepted by other students or feeling more connected to their school. These studies tend to be correlational, making it difficult to rule out alternative explanations.

The most persuasive findings come from experimental research done by Greg Walton and Geoff Cohen, two Stanford psychologists. In a series of papers, they have demonstrated that a "belonging" manipulation can lead first-year college students who, prior to the manipulation, felt like they didn't belong, to earn significantly higher grades throughout college. Specifically, they tested the effects on African-American and European-American Yale students, who were represented by 6 and 58 percent of the student body, respectively. Some students read a testimonial from an older student talking about how she had been worried about fitting in but that things had turned out really well. Others read a testimonial from an older student talking about how his political views had gotten more sophisticated over time at college (but saying nothing about fitting in). Whichever kind of testimonial the students read, they then had to deliver their own testimonial on video about the same thing.

Walton and Cohen obtained the students' GPAs during each

semester of college. For the African-American students, this single belonging manipulation led to an enduring improvement in GPA in nearly every semester of about 0.2 GPA units (for example, a GPA of 3.6 instead of 3.4). The European-American students did not show this benefit. They probably already felt like they belonged. Given how well represented they were in the student body, the manipulation wouldn't be expected to be as effective. But if this effect isn't about race per se, there might be tremendous use for it in the transition to junior high for all students, because at this juncture of their lives countless students of all races feel like they don't belong.

This is a pretty crazy finding if you think about it for a minute. Three years after spending an hour in a psychology experiment that momentarily amped up their sense of belonging, students were still benefiting from it in their academic performance. Additionally, as seniors, the students were asked about having been in the study three years earlier. Most could remember that they had been in the study, but almost none of them could remember what the study had been about. This effect was carrying on long after folks could remember that the original event had happened.

Given that social rewards make us feel good and activate the brain's reward system, these findings are consistent with past work on emotional experience and intellectual performance. Social psychologist Alice Isen repeatedly observed that feeling good ("positive affect") is associated with improved thinking and decision making. In addition, positive affect has been associated with finding similarities and differences between ideas more effectively, and two separate studies have found that positive affect enhances working memory ability.

Why would feeling good, whether it's the result of getting a surprise gift or finding out that others like you, have any impact on your ability to think well? Neuroscientist Greg Ashby has suggested the reason is that feeling good and thinking well both depend on dopamine. Whenever you do something that feels good or that is rewarding, dopamine is released from the ventral tegmental area of the brainstem and is projected to the ventral striatum. However, the ventral striatum is not the only brain region that is affected by dopamine released from the ventral tegmental area. The lateral prefrontal cortex is also rich with dopamine receptors, which means many of the cognitive functions associated with the lateral prefrontal cortex are modulated by the presence of dopamine. Dopamine reductions in the prefrontal cortex have been shown to impair working memory, and in at least some concentrations, increasing dopamine can improve working memory. Putting this all together, it is plausible that the dopamine released during feelings of social reward promotes more effective prefrontal control during classroom activities, leading to higher grades.

Social motivations—the need to avoid social pain and the need to experience social connection—are basic needs that can impair learning when unmet. Over the past two decades, some in education have seen the light about the need to address these motivations, but change is slow in coming, in part because to mentally equate social motivations with actual changes in academic performance is so hard. Yet this is only a starting point. The social brain offers other insights, which haven't been considered at all, about how to change schools to improve academic outcomes.

Thou Didst Beat Me

If our schools are broken, they have been broken in much the same way for a very long time. The very first formal classroom was established in the first century AD for Jewish children to learn the Talmud, the book of Jewish laws and customs. The classes were set up for all children above the age of six with no more than twenty-five students per class. Five rows of five? Sounds familiar. Going back even further, an Egyptian child's clay tablet from 3000 BC was inscribed with the words "Thou didst beat me and knowledge entered my head." That sounds familiar too. By junior high, education is a battle between teachers trying to get English, history, math, and science into the heads of the students, while the students are preoccupied with the stuff that is actually important to them—the immediate social world of their peers.

After our journey through the latest research on the social brain, we know that it isn't the students' fault that they are distracted by the social world. We are built to turn our attention to the social world because, in our evolutionary past, the better we understood the social environment, the better our lives became. The mentalizing system that promotes this understanding is particularly active and influential in early adolescence.

Although the brain is built for focusing on the social world, classrooms are built for focusing on nearly everything but. We spend more than 20,000 hours in classrooms before graduating from high school, and research suggests that of the things we learn in school, we retain little more than half of the knowledge just three months after initially learning it, and significantly less than half of that knowledge is accessible to us a few years later. Why do we bother with this adversarial teaching process when so little of what is learned is actually retained? Do we invest so much in our schools just so that we can say children were exposed to all of the important information? Don't we want them to actually learn and be able to use what they learn once they have finished with school?

If we want to improve our schools, we need to take a long hard look at what we are doing and be willing to toss a lot of it because it simply isn't working. If I had a printer that reproduced only 30 percent of the words I had typed on the page, I would throw it out. We need to do the same with education. I'm not one of those who think "teachers are the problem." Teachers work incredibly hard under difficult circumstances. But we are sending them to war with butter knives instead of the ammunition they need to transform our children into the adults we want them to be.

The Mentalized Classroom

Teachers are losing the education war because our adolescents are distracted by the social world. Naturally, the students don't see it that way. It wasn't their choice to get endless instruction on topics that don't seem relevant to them. They desperately want to learn, but what they want to learn about is their social world—how it works and how they can secure a place in it that will maximize their social rewards and minimize the social pain they feel. Their brains are built to feel these strong social motivations and to use the mentalizing system to help them along. Evolutionarily, the social interest of adolescents is no distraction. Rather, it is the most important thing they can learn well.

How do our schools respond to these powerful social motivations? Schools typically take the position that our social urges ought to be left at the door, outside the classroom. Talking, passing notes, or texting one's classmates during class are punishable offenses. Please turn off your social brain when you enter the classroom; we have learning to do! It's like telling someone who hasn't eaten to turn off the desire to eat. Our social hunger must also be satisfied, or it will continue to be a distraction precisely because our bodies know it is critical to our survival.

What then is the solution? Giving students a five-minute break during class to socialize? Letting them send text messages as they please? I believe the real solution is to stop making the social brain the enemy during class time and figure out how to engage the social brain as part of the learning process. We need the social brain to work for us, not against us in the learning process. Classroom learning as it typically occurs depends on the lateral prefrontal and parietal regions involved in working memory (see Figure 5.2) and reasoning, along with the hippocampus and the medial temporal lobe, which are involved in laying down new memories. As we have learned, the mentalizing system tends to operate in opposition to the traditional learning network. What we haven't discussed is that the mentalizing system can operate as a memory system too, one that is potentially more powerful than the traditional learning network.

In the 1980s, a series of behavioral studies demonstrated a curious phenomenon. In the first of these studies, social psychologist David Hamilton asked people to read statements describing ordinary behaviors (for example, "reading the newspaper"). Some of the participants were told to try to memorize all the information because there was going to be a memory test later. Other participants were told to "form an overall impression of what the person who performed these various actions is like" and were explicitly told *not* to try to memorize the information. These folks were not told about the upcoming memory test, and instead they were told they would later be asked some questions based on the impression they had formed. These individuals thought they would be asked questions like "Would this person prefer watching a movie or going hiking?" Regardless of what each participant was told to expect, everyone got a memory test. Guess who did better on the memory test: the folks told to memorize for the test, or the folks trying to form an overall impression? I wouldn't be telling you about this study if it were the memorizers, would I? As strange as it might seem, in study after study, the folks making sense of the information socially have done better on memory tests than the folks intentionally memorizing the material.

For years, this *social encoding advantage* was assumed to result from efficient use of the traditional learning system (that is, working memory regions plus the medial temporal lobe). People thought that social encoding must be using this system better than memorization attempts do. This was a parsimonious explanation, but with all due respect to William of Ockham, it was wrong. Jason Mitchell, a social neuroscientist at Harvard University, ran an fMRI version of the social encoding advantage study. As in a dozen studies before his, he found that when people were asked to memorize the information, activity in the lateral prefrontal cortex and the medial

temporal lobe predicted successful remembering of that information later on. According to the standard explanation of the social
encoding advantage, the same pattern should have been present or
even enhanced when people did the social encoding task, but that
isn't what happened. The traditional learning network wasn't sensitive to effective social encoding. Instead, the central node of the
mentalizing network, the dorsomedial prefrontal cortex, was associated with successful learning during social encoding.

The educational implications of these findings are vast. These
findings suggest that the mentalizing system is not just for social
thinking—it is also a powerful memory system. Under certain
circumstances, it appears to be a more powerful memory system
than the traditional one, as social encoding leads to better memory
performance than actually trying to memorize. This system has
limitations with respect to classroom learning, but before getting to
those, let's roll out the big news here.

The brain has a fantastic learning system that has been largely
untapped as an educational resource. This could change everything.
Remember, when we use the traditional memory system it typically suppresses the mentalizing system. That's the seesaw. Thus,
the way we have learned how to learn in the classroom turns off
this mentalizing-based learning system. Moreover, the entire structure of classroom learning is meant to prevent this system from
contributing; social thinking is penalized in the classroom. So let's
consider some simple ways to make the social brain work for us,
instead of against us, in the classroom.

History and English

I hated history class. It didn't matter whether it was U.S. history
or world history. The way it was taught and described in our text-
books was largely the history of who had power, who fought whom
in what year, which country was able to get or keep power, and

what the territorial boundaries looked like before and after each war. The facts as currently taught are commonly devoid of the social content and implications that the mentalizing network seems to naturally crave, leading minds to wander to other distractions that impede learning. Yet historical events, as they are occurring, are intensely social, nearly always infused with multiple mentalistic narratives that we use to understand the events while they are still in the headlines.

Consider the ongoing diplomatic standoff between the United States and Iran. U.S. leaders are constantly debating the true beliefs of Iranian leaders regarding their goals for uranium enrichment. Do they want it as a power source as they claim or to create nuclear weapons? Most U.S. policy wonks assume Iranian leaders are not telling the truth, but what can be inferred about their true intentions from the things they do say? Emotionally, U.S. leaders are driven by their fears about the implications of a nuclear Iran and how that would affect Israel, our closest ally in the Middle East. Strategically, Democrats and Republicans are jockeying for position on the rhetoric of our Iran policy in order to weaken or embarrass their opponents in the next election. All of these events are rich with social and mentalistic drama, but by the time they become a paragraph in the history textbook, this social cognitive drama will all be stripped away and replaced with the actions that were ultimately taken.

It is understandable that historians would worry about sticking to the objective facts and shy away from the kinds of inferences needed to discuss the psychosocial drama of historical events. But as someone who cares about education, I care far more about how to make this content interesting to students. History has always been a giant soap opera as it unfolds, which is part of why current events are so interesting. They appeal to our mentalizing system, which wants to understand the *why* behind all of the observable events. News analysis provides us with this mentalistic drama, and if history classes are to be more engaging, they should as well. History class needs to move from limiting discussion to the *how* and *what* of history to the much richer *why* that students crave. Presenting historical events in terms of possible social narratives that address the thinking, feeling, and motivations of historical figures may well improve retention of the agreed-upon key facts by invoking the mentalizing-based memory system. It is like hiding medicine inside a piece of candy; the child enjoys the candy while the candy also serves as a vehicle for the medicine.

Social thinking is equally relevant to English class and yet is largely absent from the curriculum. English classes devote enormous time to the rules of writing. Lessons focus on spelling, grammar, syntax, topic sentences, and the five-paragraph paper. These are commonly presented as a rigid set of facts and rules to be learned and implemented in one's writing. The true motivator behind all of these facts and rules lurks in the shadows, rarely discussed openly in the classroom: good writing is all about getting ideas from your mind into the minds of other people, so that they understand you and are persuaded, informed, or moved by you. This is a straightforward mentalizing concept that should be students' North Star, their guiding light.

Based on social brain research, I would argue that rather than English classes, students should be taking "communication classes" because this would place the focus on all the tools we have at our disposal to communicate effectively with others, if only we learn how to wield them. Understanding the minds of one's audience and how they are likely to interpret or misinterpret what has been written is the essential principle behind the rules of good writing. Consider the use of the passive voice. Everyone learns that using the passive voice is a major no-no (for example, "the bicycle was ridden by the boy" should be "the boy rode the bicycle"), but few learn the reason for this: the reader has to do more mental work to understand passive language. The passive voice isn't wrong because it violates a sacred principle. It is wrong, in most cases, because it is harder to follow. The deeper principle is "Make yourself easy to understand." This is the litmus test for nearly all writing decisions. While history class would be improved by focusing on *why* historical figures did *what* they did, English class would be equally improved by focusing on *why* each rule improves comprehension and *when* it does so, rather than exclusively focusing on *what* the rules are. Our brains crave *why* stories, and in history and English, these are natural companions to what is already taught.

Math and Science

When policy makers sound the alarm about how U.S. classrooms are falling behind those in other countries, they aren't thinking about history class, and most of them aren't focused on English either. There is a huge federal initiative to improve education in the so-called STEM fields (science, technology, engineering, and math) because these fields are seen as determining our ability to improve the quality of life through new inventions, techniques, and discoveries. As much as I am trying to shed light on the overlooked significance of the social world and how our brain specifically evolved to interact with it, I am not making the case that mentalizing has a natural home in lectures on geometry or organic chemistry. In these subjects, the social brain may not play a role in the content of the material. However, it may still end up being central to the learning process. If we can increase the social motivations present while learning math and science, students may absorb the content far better than they currently do on average.

Recall the social encoding advantage: when we encode information socially, the social brain handles the encoding and leads to better retention of the information than the traditional memory system. Given that it makes no sense to ask people to think about math from a social perspective, the trick is to get the mentalizing system to do the learning a different way—to get learners to think of themselves as teachers while they are learning. Yale psychologist John Bargh first performed this trick back in 1980.

In the first study of *learning-for-teaching*, Bargh compared people who were memorizing information for a test to people who had learned the same material in order to be able to teach it to someone else. Those learning the material in order to teach it performed better on what to them was a surprise memory test, compared to those who knew the memory test was coming and had studied for it. There are two other key facts of this study. First, those learning in order to teach never actually got to teach. The act of teaching no doubt enhances memory; however, in Bargh's study, people were tested right after learning the material. Thus, any advantage they showed had to be due to the social motivation present while learning the material. The second critical fact is that the material to be learned was not social. This means that the results of this study might generalize to learning a variety of nonsocial content—like science and math.

It is an open question as to whether the learning-for-teaching effect is like the social encoding advantage, in terms of using the mentalizing system rather than the traditional memory system. My lab is currently investigating this, but we do already have some suggestive evidence that social motivation alone is sufficient to activate the mentalizing system's memory abilities.

Recall that in Chapter 5 on mentalizing, I described a study that Emily Falk and I ran looking at what happens in the brain when we see messages that we are destined to spread effectively to others—when we are acting like Information DJs. One of the things we looked at in our analyses was how accurately the Interns (the people in the scanner who were the first to be exposed to the pilot television show descriptions) remembered the ideas when passing the information on to the Producers (the people who learn about the pilot shows only from the Interns). Memory accuracy was associated almost exclusively with mentalizing activity, not the traditional memory system, when the idea was first presented. This suggests that the motivation to share the information may have been routing the content through the mentalizing system for later retrieval.

If we know that social motivations to learn can invoke the mentalizing system and enhance learning, how do we apply that knowledge in classes such as math and science? Perhaps through peer tutoring—students teaching students. Rather than trying to prevent student interactions during class, the view from the social brain is that such talking should be encouraged, but focused, to maximize the benefits. Peer tutoring has been used in classrooms, but not broadly and not in ways that maximize its social motivational benefits for learning.

Consistent with the learning-for-teaching findings, multiple studies have demonstrated that peer tutoring benefits learning in both tutors and tutees, with tutors often benefiting more. In some ways this has been seen as a limitation of peer tutoring because the intention has been to specifically enhance tutee learning and close the gap between low- and high-achieving individuals. From the current perspective, a broad program focused more on *tutor* learning (through socially motivated teaching), with all students functioning as both tutors and tutees, might lead to the best educational outcomes. We should figure out ways to make low achieving students the tutors, where the learning benefits are greatest, rather than always putting them in the role of the person being tutored.

Consider how peer tutoring would be experienced from the vantage point of a typical eighth grader. Instead of receiving a forty-minute lecture from a teacher, each eighth grader would spend twenty minutes teaching a sixth grader about lowest common denominators and then would receive a twenty-minute lesson from a tenth grader about basic algebraic equations. Of course, the teacher would have to work with the eighth graders to make sure they were ready to teach the sixth-grade lesson. However, the eighth graders, by also working with the sixth and tenth graders, would have multiple social motivations to learn the material.

Typical eighth graders may not enjoy listening to a teacher teaching them about math, but I bet they will care substantially about the teaching relationship they have with the sixth graders. Junior high students want to be in charge and feel autonomous, and learning-for-teaching would provide a great opportunity. It is a chance for them to be an authority in the eyes of the adoring sixth graders, who often will leap at the chance to spend time with the cool eighth graders. Recall from Chapter 11 that knowing that one's efforts are really helping someone leads to more and better effort. To make personal learning about helping others would set these motivational processes in motion. Older students often like having a few younger students they look out for, and peer tutoring offers an educationally focused way to accomplish this.

Naturally, there is also the fear of screwing up in front of the sixth graders, which motivates the eighth graders to learn the lesson as well. Social embarrassment can be a stronger motivator than the fear of a low score on a test. All of these social dynamics repeat themselves when the tenth graders teach the eighth graders. The eighth graders may not like listening to adults, but what could be better than hanging out with the tenth graders? Finally, I also suspect that eighth graders will like their adult teachers more if they feel they are actively collaborating with them in order to teach the sixth graders.

Remember that the trick in all this is that when the eighth graders learn the lesson to teach the sixth graders, the eighth graders will be more likely to engage the mentalizing system to boost the quality of the memory for the material. The potential downside is that the same material will be taught twice to each student. First, time will be spent having the eighth graders teach the sixth graders a lesson, and then those sixth graders will have to spend time revisiting the material two years later in eighth grade in order to teach it to a new group of sixth graders. By definition, students will cover less content in school. But given that students are remembering so little of what they are learning now, wouldn't it be better to teach them less and have them actually learn and retain the material?

Social Brain Class

While we are taking advantage of the fact that most class time is spent on material that will be forgotten, perhaps we should use some of that wasted time to learn something else. Our brain craves to understand itself, the social world, and the relation between the two. This understanding is what the mentalizing system and the self-processing regions make possible. Neural and hormonal changes during adolescence make this goal even more pressing. Why not spend at least part of the day teaching what the brain is most biologically prepared to learn about? Effective social skills are as important to most careers as other facts and analytical skills currently being taught in school. Being able to work effectively with team members, superiors, and subordinates is critical to most work environments. Can anyone make the argument that algebra is as important as social intelligence to most people's professional or personal development? Do you believe that everyone around you already has as much social intelligence as they need?

Despite the regular practice that the mentalizing network gets from birth to adulthood, our social expertise is clearly less than it could be. Unlike nearly everything else in life, we are each left to our own devices to figure out the social world. If you want to play piano or play soccer, you take a class or get a coach who corrects you each time you make an error. But for learning about the social world, you are on your own. We rarely get direct feedback about our errors in our social thinking. Partly for this reason, people are susceptible to a wide variety of social cognitive and self-processing errors and biases, including but not limited to naïve realism, fundamental attribution errors, false consensus effects, affective forecasting errors, in-group favoritism, and overconfidence. How can you correct these errors if you never have them pointed out when you make them? How do you learn what the correct inference should have been, and why? Teaching our students about these processes—why they occur and how to identify when we are making these errors—won't eliminate all of them, though it would likely diminish some. What it will do is provide a shared language for discussing and considering these errors when they occur, which in turn will help people understand that the errors that others make usually aren't malicious or intentionally self-serving. No one wakes up in the morning and says, "I've got to work harder at being a jerk today." We all make frequent social mistakes, and I suspect we always will. If we, and those around us, had a more mature understanding of them, we would be able to stop these errors in their tracks and minimize the fallout that comes from their being misinterpreted.

We should be teaching our students about their social motivations and the fact that hurting someone else's feelings is more like a physical assault than we might intuitively believe. We should teach our students that it is natural to have both selfish and prosocial motivations and that the latter do not need to be hidden. We should teach our students that our desires to be socially connected aren't a weakness and that our interest in understanding the social world is an evolutionary advantage that has been baked into our operating systems over millions of years. The developing social brain needs accurate information about the social world. Too many of our adolescents are getting these models from sitcoms and uninformed peer opinion. There is a science of how the social world works, and social psychology, social neuroscience, and sociology all have a lot to tell us about it. We have an opportunity to craft far more socially savvy adults. And there is little doubt that teachers would have less difficulty keeping their students' attention in this particular class; it's exactly what the adolescent brain craves.

Exercise Class for the Social Brain

The film *Gattaca* explores the commonly held notion of genetic determinism taken to its frightening extreme. Everyone is born from a test tube in which only the best DNA from each parent was used to create the most nearly perfect person possible. The central theme of the movie examines whether individuals born genetically "inferior" can overcome their limitations through hard work. The premise is plausible because most of us hold contradictory beliefs that we trot out at different times, depending on what suits us. On the one hand, we tend to believe the cards we are dealt genetically, at birth, are hugely determinative of the kind of life we will have. On the other hand, we subscribe to the view that through hard work we can get further in life than we would otherwise.

The determinist view was buffeted for a long time by the belief that the human brain was relatively fixed and had all the neurons it was ever going to have not long after birth. If we think of the brain as a computer, this viewpoint leads to the ancillary belief that we can change the contents of our hard drives (that is, learn new information) but not change how the hard drive works (that is, the processes that support thinking and learning). It is perhaps for this reason that education is so focused on the acquisition of new information rather than on trying to mold minds themselves (despite occasional claims to the contrary).

Times have changed, and it is clear to neuroscientists, if not yet to the rest of the world, that the neuronal makeup of our brains is far more flexible than previously believed. Neuroscientist Liz Gould discovered that new neurons can be born in adulthood, and this process might be stimulated by exercise. People who were learning to juggle for a few short months came to have greater cortical thickness in brain regions involved in motion perception, an effect that lasted long after the individuals ceased juggling. Similarly, taxi drivers in London, who must learn an extremely complex map of streets, come to have larger hippocampal volume the longer they have been working.

The fact that our brains are more malleable than previously assumed has led scientists to start focusing on the kinds of experiences that can change how the brain works. One of the most exciting lines of research has focused on working memory training. Although working memory capacity and fluid intelligence were long thought to be fixed traits, a series of recent studies have shown that working memory training can alter both working memory and fluid intelligence with concomitant neural changes.

Is working memory just the tip of the iceberg? Could we train our brains to be better at mentalizing, empathizing, and exerting self-control? There is no question that these are unqualified assets and that more of each in our society would be a good thing. While teaching about the social brain is a great idea, having "exercise class" for the social brain might be an even better one. For twenty minutes a day, seventh and eighth graders could do various kinds of training exercises to strengthen and fine-tune the social brain. How much of your early education would you trade to be able to better read the minds of others and to be more able to overcome your impulses?

According to recent work by social neuroscientists Jen Silvers and Kevin Ochsner, emotionality peaks right around eighth grade, but our capacity for emotion regulation doesn't reach its full maturity until we are exiting our teenage years. Besides making teens difficult to deal with at times, this hyper-emotionality can also put them at a significantly higher risk of making life-altering bad decisions leading to delinquency, addiction, pregnancy, and dropping out of school. If a social brain exercise class could change those trends and give our students greater psychological resources for focusing in class, doing homework, and studying for tests, this would be profoundly beneficial.

How do we train self-control? In Chapter 9, we saw that incredibly diverse forms of self-control all seem to rely on the right ventrolateral prefrontal cortex. From delay of gratification and emotion regulation, to perspective taking and overcoming motor impulses, success is nearly always associated with activating this region. Do we need to train each of those different kinds of self-control separately? There is growing evidence that practicing self-control in one of these domains can pay dividends in the others.

In one study in my lab, Elliot Berkman and I examined the effects of training motor self-control on the ability to regulate one's emotions. Individuals were either assigned to the self-control training group or the non-self-control group. Non-self-control group members came to our lab eight times over two to three weeks to practice a very simple visual-motor task. Left- and right-facing arrows were presented on the screen one after another, and the corresponding arrow key on the keyboard needed to be pressed as quickly as possible. The only self-control needed for the non-self-control group was overcoming the urge not to return to the lab again to spend more time on this boring task. Those in the self-control group worked on a self-control variant of this task each time (the stop-signal task described in Chapter 3). Every so often, a tone would sound after one of the arrows appeared, and this sound would indicate that the person should not press a button at all that time.

Regardless of condition, all of the participants in the study came for an initial testing session, during which we measured their emotion regulation ability using a reappraisal task. Individuals were shown aversive images. Sometimes the participants were asked to fully experience their emotional reaction to the images, but on other trials they were asked to reappraise the images, thinking about them in a way that would make them less distressing. By comparing how distressing the images were for participants under these two instructions, we were able to compute a measure of how well individuals could regulate their emotional responses using reappraisal. Three weeks later, after completing all the practice sessions with the visual-motor task, the participants returned and had their emotion regulation ability tested one more time. Critically, participants did not engage in any kind of emotion regulation training in the interim.

What we were interested in discovering was the effect of visual-motor self-control training on emotion regulation ability—even though these two things seem to have little in common. Indeed, there was a relationship for those in the training group. Individuals who had received self-control training with the visual-motor task had significantly better emotion regulation ability at the end of the study than they had had at the beginning, even though there was no emotion regulation training during the study. To examine whether motor self-control could have been driving this effect, we looked at the relationship between motor self-control improvements and emotion regulation improvements. The better an individual got at motor self-control over the course of the eight training sessions, the more their emotion regulation ability improved.

The implication of these findings for schools is that we could present students with a wide variety of ways to practice self-control, and many of them would likely create benefits across an array of self-control domains. In fact, mindfulness meditation may turn out to be a great way to strengthen this muscle, as it has been shown to enhance ventrolateral prefrontal responses. There are many ways to exercise this self-control muscle, and doing so can help our self-control efforts throughout our lives.

Each proposal I've suggested here requires hard choices to be made. There is only so much time, money, and human energy to put toward education; giving these resources to one initiative means diverting them from another. But it is worth doing the hard work to get this right. Every junior high student who stays engaged rather than losing interest is vastly more likely to go to college and make greater contributions to their community. It is natural to believe that education should be primarily about presenting the most important facts to children and expecting them to absorb and retain them. But education doesn't work that way. The smartest kids with the strongest natural self-control can force themselves to learn this way, but the great majority of our students can't. Shaping the context and curriculum in light of what we are learning about the social brain will help our students maximize their potential. Over the last generation, we have turned all of our C students into B students through grade inflation. Wouldn't it be great if we could turn all those B students into A students because they learned more?

It is useless to attempt to reason a man out of a thing he was never reasoned into.

—JONATHAN SWIFT

The great double-edged sword of being the most sophisticated mammals on the planet is that no matter how smart or rational we become, we can't outthink our basic needs. We all need people to love and respect, and we all need people who love and respect us. Would life without them be worth it? Does the ability to play chess and solve calculus problems make up for a life without other people? Mother Teresa, who observed people in the most squalid living conditions imaginable, believed that a life without other people "is the worst disease that any human being can ever experience." Those basic social needs are present at birth to ensure our survival, but we are guided by these needs until the end of our days. We do not always recognize these needs, and we may not see them influencing those around us, but they are still there nonetheless.

Our basic urges include the need to belong, right along with the need for food and water. Our pain and pleasure systems do not merely respond to sensory inputs that can produce physical harm and reward. They are also exquisitely tuned to the sweet and bitter tastes delivered from the social world—a world of *connection* and threat to connection. A condescending look from a complete stranger can feel like a dagger, just as a kind look can reassure us that we are safe in a new environment. As we saw, evolutionarily, this wiring came from the need to keep mammalian young, who are born too immature to fend for themselves, close to their caregivers. The dorsal anterior cingulate cortex and the anterior insula treat real or potential separation from others as painful, which promotes distress calls in mammalian young when separated from their mothers. In contrast, the reward system is sensitive to both giving and receiving care, which promotes social bonding from the perspective of parent and child, respectively. When we experience social pains or feel the distress of withheld social connection, we are unable to focus on much else until this need is met.

Social pain and pleasure make use of the same neural machinery as physical pain and pleasure, creating a powerful motivational drive to maximize our positive social experiences and minimize our negative ones. Luckily, evolution has given us an arsenal of social weapons to help satisfy these social needs and ensure group cohesion. Emerging to some degree in other primates, the capacity for *mindreading* allows us to consider the goals, intentions, emotions, and beliefs of others. Monkeys see the world in terms of the psychologically meaningful actions and expressions of others thanks to the mirror system. This allows them to empathize, help others, and coordinate their activity in many situations. In humans, our social imagination, processed via the mentalizing system, primarily in the dorsomedial prefrontal cortex and the temporoparietal junction, allows us to take this coordination to the extreme, creating various symbolic social connections like the attachment we feel to sports teams, political parties, and even celebrities. Fortunately, it also allows us to build societal institutions for government, education, and industry. And it allows us to take deep pleasure in fiction, whether in books, on the big screen, or on the small screen. Indeed, most pleasures that we experience beyond sex and drugs (and even those to a degree) depend on our ability to imagine the experiences of others.

The mentalizing system is coherently active from the moment we are born, reliably comes on whenever we have downtime, and is even on while we dream. The brain is designed to devote as much time as possible during our development and adulthood to mentalizing activities. We don't know exactly what the mentalizing system is doing during rest (because as soon as you ask someone, that person isn't at rest anymore). But we do know that people who turn the mentalizing system on more at rest also do better in general at understanding the minds of others. People who happen to have the mentalizing system spontaneously activated seconds before doing a mentalizing task do the task better. These results suggest that at rest the mentalizing system may be rehearsing and reconsolidating various kinds of social information that will benefit our long-term social abilities. And in the moment, it may nudge us to see the world through a social lens. For those of us who wondered whether our sociality is just an accident, this is strong medicine: our brain sets itself to see the world socially, presumably because of the great benefits of doing so.

Finally, we humans have the ability to reflect on ourselves, to think about our characteristics, beliefs, and values in relation to other people, and then to deploy our capacity for self-control to restrain undesired impulses in order to pursue our long-term goals. As we have seen, the way the self-architecture is set up in the medial prefrontal cortex and the right ventrolateral prefrontal cortex, it also serves a more duplicitous purpose that we are generally unaware of in daily life. The self system serves as a Trojan horse, sneaking in the values and beliefs of those around us under the cover of night without our ever being the wiser. So when we use our capacity for self-control to pursue our goals and values, they are quite often goals and values that will benefit society as much as or more than they will benefit us personally. And when we are made aware of ourselves as social entities that can be judged by others, our self-control often kicks in to ensure that we act in accordance with the values of those around us. A self system that operates this way improves our odds of being liked, loved, and respected by members of the groups we are in because we will work hard in pursuit of the groups' goals and values. These mechanisms are the glue that keep us *harmonizing* with one another much more easily than we otherwise would.

When I started to think about writing this book, I thought there was a series of pretty cool findings in social cognitive neuroscience worth sharing. I thought each of them stood on its own, independent of one another. Today my perspective is quite different. I see a tapestry of neural systems woven together to bind us to one another. Our social brain keeps expanding, using the existing building blocks to further enhance our ability and inclination to be social. The fact that we like soap operas, reality TV, and gossip isn't a strange accident of having a complex mind. It's the natural consequence of having brains that were built to make sense of other brains and to understand everyone's place in the pecking order.

As exciting as the last twenty years have been in terms of laying the groundwork for understanding the social brain, the next twenty may prove to be all the more exciting. Increasingly, neuroimaging techniques will allow us to measure the neural bases of people's social and emotional experiences as they unfold in real-world social interactions. Functional near infrared spectroscopy (fNIRS) is a technology that shows real promise in this regard. fNIRS subjects can essentially just wear a headband that has light emitters that direct light through the skull. As the light hits brain tissue, it scatters back out of the brain in such a way that we can detect when a brain region is more or less active. How it works is inside baseball, and it has many limitations of its own, but the essential thing is that it allows fMRI-like studies to be conducted while a person is sitting up, talking, and interacting with one or more other people. fMRI requires the participant to lie down, alone, on a surgical bed slid into a giant mechanical donut. In contrast, fNIRS can be conducted wirelessly, allowing two people, each wearing an fNIRS headband, to take a stroll together while their neural activity is being transmitted back to a base station. And whereas MRI scanners can cost $3 million and another $1 million to install, an fNIRS headband can be purchased for under $100,000, meaning that this equipment could be broadly accessible in schools, businesses, or even psychotherapy offices as the price comes down gradually over time. As more and more groups study the social brain in more and more real-world contexts, we will gain increasing perspective on how the mind works when it is fully immersed socially.

In Isaac Asimov's science fiction classic *Foundation*, Hari Seldon creates a new branch of mathematics called "psychohistory," in which principles of psychology are used to predict how major geopolitical events over the next several decades will materialize and be resolved. As sinister as such a tool might seem in the wrong hands, it could also allow us to have unprecedented quality of life. We are fundamentally psychological creatures, social psychological creatures. Stock markets are moved as much by our general hopes and fears as they are by the fundamentals and specific activity of the any stock. As we learn more about our sociality from psychology, neuroscience, and beyond, we have a great opportunity to reshape our society and its institutions to maximize our own potential, both as individuals and together as a society. Someday the president will consult with social neuroscientists and psychologists when making policy decisions. Someday, CNN will want to add experts on the social mind and social brain to their stable of political scientists, political strategists, and economists when making sense of world events. Someday we will look back and wonder how we ever had lives, work, and schools that weren't guided by the principles of the social brain. The years ahead that will change this from science fiction to science will be exciting ones indeed.

ACKNOWLEDGMENTS

Finally, I get to say thank you to all the people in my social network who helped make this book possible. Amazingly, as I count up all the folks I am grateful to, it comes awfully close to Dunbar's number. My family has been supportive since the beginning (I'm talking diapers)—thanks to the Liebermans, Albucks, and Eisenbergers for your encouragement over the years.

The first intellects who lit a fire for me were philosophers such as Nietzsche and Sartre. They have influenced my thinking since I was a teenager (and thanks to Dad for having been a philosophy major and leaving those books on the shelf even if they weren't his favorites). In college, my philosophy mentor, Bruce Wilshire, convinced me that the *big problem* to understand was "experience, hot off the griddle." Though neuroscience strays far from this dictum, it has stayed with me as something to strive for.

In many ways, I owe my career to three people from Harvard. My PhD advisor, Dan Gilbert, never ceases to inspire me to be a better social psychologist (as amazing as his writing is, he's even better in person). Steve Kosslyn oversaw an undergraduate course that I taught and switched the curriculum from focusing on Freud, Skinner, and cognitive science to one in which each topic was taught from three levels of analysis: social, cognitive, and neuroscientific. Kevin Ochsner was a fellow graduate student, and we took the initial steps of this journey into social cognitive neuroscience together, in the process becoming best friends and each other's best man. I am so grateful for how these three people shaped my thinking and my research at such a pivotal moment in my career. Without them, there is no way I could have written this book.

Since arriving at UCLA almost fifteen years ago, I have been able to do the work described here because of the outstanding collaborators and mentors I've had: Lori Altshuler, Susan Bookheimer, Ty Cannon, Mark Cohen, Michelle Craske, Mirella Dapretto, Alan Fiske, Andrew Fuligni, Adam Galinsky, Ahmad Hariri, Marco Iacoboni, Michael Irwin, Edythe London, Emeran Mayer, John Mazziotta, Bruce Naliboff, Annette Stanton, Shelley Taylor, and Kip Williams. I'm grateful to have been able to work with each of you. I'm also thankful to have spent time with Scott Gerwehr, Bryan Gabbard, and everyone at DGI who helped me think about how social neuroscience could be applied in the real world. Scott, we still miss you.

I have also been blessed with the best students and staff in the world. I'm so grateful that I've been able to spend my time thinking with everyone who has come through the UCLA Social Cognitive Neuroscience Laboratory: David Amodio, Elliot Berkman, Lisa Burklund, Liz Castle, Joan Chiao, Jessica Cohen, David Creswell, Molly Crockett, Janine Dutcher, Emily Falk, Ben Gunter, Kate Haltom, Erica Hornstein, Tristen Inagaki, Johanna Jarcho, Yoona Kang, Carrie Masten, Sarah Master, Meghan Meyer, Mona Moieni, Sylvia Morelli, Keely Muscatell, Junko Obayashi, Jenn Pfeifer, Josh Poore, Lian Rameson, Ajay Satpute, Julie Smurda, Bob Spunt, Golnaz Tabibnia, Eva Telzer, Sabrina Tom, Jared Torre, Stephanie Vezich, Baldwin Way, Locke Welborn, and Charlene Wu. All of you are amazing.

Without my agent, Max Brockman, *this* book would never have existed. He played the role of Socrates' Daimon for a long time. Thank you for saying no to the first several proposals I sent you over the course of two years and then finally saying yes when we got to this one. As frustrating as those no's were at the time, I shudder to think of what I would have ended up with had you said yes to any of my initial proposals.

To my editor, Roger Scholl, and everyone at Crown—thank you so much for your help and support, every step of the way. You have been fantastic to work with. You made something incredibly daunting into something merely daunting.

As I began writing the book itself, I made a point of spending some time reading fiction every day to try to keep my mind nudged toward writing more like a novelist and a little less like an academic. I read wonderful books by Haruki Murakami, Hugh Howey, Paul Auster, Michael Cox, David Mitchell, Kazuo Ishiguro, Philip K. Dick, Matthew Mather, Ernest Cline, and Wilkie Collins. Along similar lines, I'm glad I had Pandora on every day while I was writing with Tycho, Ulrich Schnauss, Riceboy Sleeps, Ambulance, William Orbit, Vector Lovers, Loess, Casino versus Japan, Brokenkites, Her Space Holiday, Deosil, Infinite Scale, Boards of Canada, Trentemoller, Eluvium, William Basinski, and Michael Maricle all in the mix. All of you helped keep in me in the right frame of mind—absorbing the part of my mind that needed some level of constant distraction.

I'm grateful to my colleagues and friends who gave me feedback on particular chapters: Robin Dunbar, Sarah Endo, Dan Gilbert, Jonah Lehrer, Jenn Pfeifer, Eva Telzer, Nim Tottenham, and James Yang. Special thanks go to the Neuroleadership Institute. For years, you've allowed me to give talks to people who care about applying neuroscience findings in organizational and educational contexts, and this experience has shaped my thinking again and again. In particular, I want to thank the folks who heard my plea for chapter reviews and went on to give me awesome feedback: Samad Aidane, Tom Battye, Marcy Beck, Pratt Bennet, Ken Buch, Corinne Canter, Christine Comaford, Garry Davis, Jon Downes, Barrie Dubois, Mary Federico, Sara Ford, Todd Gailun, Philip Greenwood, Robert Hutter, Shelley Johnson, Kory Kogon, Per Kristiansen, Kate Larsen, Dan Marshall, Jason Ollander-Krane, Bert Overlack, Thaler Pekar,

Lynn Quinn, Al Ringleb, Lisa Rubinstein, Sylvan Schulz, Mary Spatz, Bonnie St. John, Robert Weinberg, Lucy West, Scott Winter, and Susan Wright.

I want to single out two people for high praise—each of whom was involved in more ways than I can count in bringing *Social* into existence (though neither is to blame for any of its imperfections). First, David Rock, director of the Neuroleadership Institute. Since our first (dodgy) meeting several years ago, you've pushed me to make the neuroscience relevant to non-neuroscientists and to work harder to learn to communicate better so I could do that. Thank you for all the opportunities to connect at your summits with all the wonderful folks in the NLI world. It has been transformative. Thank you for reading the entire book and giving me line-by-line feedback. Thank you for spending that day in San Francisco going over all the feedback and helping me brainstorm changes for the book. Back when the book was only ten chapters, you wrote at the end: "nine chapters smart, one chapter useful!" You challenged me to write new chapters on well-being and the workplace after I thought I had written the last word, and I believe the book is much better for it.

Finally, to Naomi. There would be no book to write without you, and I certainly would not have been capable of writing it. Since I came to UCLA, you have been my partner in work and in life. Every bit of my science and every chapter of this book is better because of you. And all the feedback you have given me on my writing has always made it better—every single time. You are brilliant, and you amaze me again and again (and again). I appreciate everything you have done for me so deeply—for this book, through our research at UCLA, and in our life together with Ian.

....

NOTES

Chapter 1: Who Are We?

4   **The research my wife and I have done** Eisenberger, N. I., Lieberman, M. D., & Williams, K. D. (2003). Does rejection hurt? An fMRI study of social exclusion. *Science, 302,* 290–292.

6   **Mondale, not exactly a spring chicken** Banville, Lee. (2002). "Former Vice President Walter Mondale (Democrat)." *Online NewsHour.* PBS. Retrieved March 26, 2011.

6   **Social psychologist Steve Fein asked** Fein, S., Goethals, G. R., & Kugler, M. B. (2007). Social influence on political judgments: The case of presidential debates. *Political Psychology, 28*(2), 165–192.

7   **Imagine watching the debate yourself** Pronin, E., Lin, D. Y., & Ross, L. (2002). The bias blind spot: Perceptions of bias in self versus others. *Personality and Social Psychology Bulletin, 28*(3), 369–381.

7   **While we tend to think it is our capacity** Dunbar, R. I. M. (1998). The social brain hypothesis. *Evolutionary Anthropology, 6,* 178–190

8   **In many situations, the more you turn on the brain network** Fox, M. D., Snyder, A. Z., Vincent, J. L., Corbetta, M., Van Essen, D. C., & Raichle, M. E. (2005). The human brain is intrinsically organized into dynamic, anticorrelated functional networks. *Proceedings of the National Academy of Sciences of the United States of America, 102*(27), 9673–9678.

11   **In the toddler years, forms of social thinking develop** Herrmann, E., Call, J., Hernández-Lloreda, M. V., Hare, B., & Tomasello, M. (2007). Humans have evolved specialized skills of social cognition: The cultural intelligence hypothesis. *Science, 317*(5843), 1360–1366.

12   **During the preteen and teenage years, adolescents** Costanzo, P. R., & Shaw, M. E. (1966). Conformity as a function of age level. *Child Development,* 967–975.

Chapter 2: The Brain's Passion

15 **In 1997, Gordon Shulman and his colleagues** Shulman, G. L., Corbetta, M., Buckner, R. L., Fiez, J. A., Miezin, F. M., Raichle, M. E., & Petersen, S. E. (1997). Common blood flow changes across visual tasks: I. Increases in subcortical structures and cerebellum but not in nonvisual cortex. *Journal of Cognitive Neuroscience, 9*(5), 624–647; Shulman, G. L., Fiez, J. A., Corbetta, M., Buckner, R. L., Miezin, F. M., Raichle, M. E., & Petersen, S. E. (1997). Common blood flow changes across visual tasks: II. Decreases in cerebral cortex. *Journal of Cognitive Neuroscience, 9*(5), 648–663.

17 **The early name given to describe the network** Mckiernan, K. A., Kaufman, J. N., Kucera-Thompson, J., & Binder, J. R. (2003). A parametric manipulation of factors affecting task-induced deactivation in functional neuroimaging. *Journal of Cognitive Neuroscience, 15*(3), 394–408.

18 **The second name given to this network** Raichle, M. E., MacLeod, A. M., Snyder, A. Z., Powers, W. J., Gusnard, D. A., & Shulman, G. L. (2001). A default mode of brain function. *Proceedings of the National Academy of Sciences, 98*(2), 676–682.

19 **the network in the brain that reliably shows up** The story isn't quite this simple. There is a small subnetwork of the default network that does not typically show up in studies of social cognition, but the vast majority of the two networks are overlapping.

20 **One study looked at which brain regions were engaged** Gao, W., Zhu, H., Giovanello, K. S., Smith, J. K., Shen, D., Gilmore, J. H., & Lin, W. (2009). Evidence on the emergence of the brain's default network from 2-week-old to 2-year-old healthy pediatric subjects. *Proceedings of the National Academy of Sciences, 106*(16), 6790–6795; Smyser, C. D., Inder, T. E., Shimony, J. S., Hill, J. E., Degnan, A. J., Snyder, A. Z., & Neil, J. J. (2010). Longitudinal analysis of neural network development in preterm infants. *Cerebral Cortex, 20*(12), 2852–2862.

20 **the claim that Malcolm Gladwell made famous** Gladwell, M. (2008). *Outliers: The Story of Success*. New York: Little, Brown; Anders Ericsson, K. (2008). Deliberate practice and acquisition of expert performance: A general overview. *Academic Emergency Medicine, 15*(11), 988–994.

20 **One study found that 70 percent of the content** Dunbar, R. I., Marriott, A., & Duncan, N. D. (1997). Human conversational behavior. *Human Nature, 8*(3), 231–246.

21 **when Robert Spunt, Meghan Meyer, and I gave people only a few seconds** Spunt, R. P., Meyer, M. L., & Lieberman, M. D. (under review). Social by default: Brain activity at rest facilitates social cognition; Buckner, R. L., Andrews-Hanna, J. R., & Schacter, D. L. (2008). The brain's default network. *Annals of the New York Academy of Sciences, 1124*(1), 1–38.

21   **when you read the word "face"** Rubin, E. (1915/1958). Figure and ground. In D. C. Beardslee & M. Wertheimer (Eds.). *Readings in Perception.* Princeton: NJ: Van Nostrand, pp. 194–203.

21   **You are more likely to see faces at first** Agafonov, A. I. (2010). Priming effect as a result of the nonconscious activity of consciousness. *Journal of Russian and East European Psychology, 48*(3), 17–32.

24   **"general intelligence applied to social situations"** Wechsler, David (1958). *The Measurement and Appraisal of Adult Intelligence,* 4th ed. Baltimore: Williams & Wilkins, p. 75.

24   **In a study of more than 13,000 people** Vitale, S., Cotch, M. F., & Sperduto, R. D. (2006). Prevalence of visual impairment in the United States. *JAMA: Journal of the American Medical Association, 295*(18), 2158–2163.

24   **can we conclude our sociality is an accident** Stravynski, A., & Boyer, R. (2001). Loneliness in relation to suicide ideation and parasuicide: A population-wide study. *Suicide and Life-Threatening Behavior, 31*(1), 32–40.

24   **Friendship has been documented in** Silk, J. B. (2002). Using the "F"-word in primatology. *Behaviour,* 421–446.

25   **the closer friends become, the less they tend to keep** Fiske, A. P. (1991). *Structures of Social Life: The Four Elementary Forms of Human Relations: Communal Sharing, Authority Ranking, Equality Matching, Market Pricing.* New York: Free Press.

25   **Americans spend 84 billion minutes per month** Bureau of Labor Statistics: http://www.bls.gov/home.htm.

25   **we give an average of $300 billion a year** "U.S. charitable giving approaches $300 billion in 2011": http://www.reuters.com/article/2012/06/19/us -usa-charity-idUSBRE85I05T20120619.

26   **The brain regions reliably associated with general intelligence** Fox, M. D., Snyder, A. Z., Vincent, J. L., Corbetta, M., Van Essen, D. C., & Raichle, M. E. (2005). The human brain is intrinsically organized into dynamic, anticorrelated functional networks. *Proceedings of the National Academy of Sciences of the United States of America, 102*(27), 9673–9678.

26   **the more this network turns on, the more the general** Van Overwalle, F. (2011). A dissociation between social mentalizing and general reasoning. *NeuroImage, 54*(2), 1589–1599.

26   **To the extent that the social cognition network stays** Anticevic, A., Repovs, G., Shulman, G. L., & Barch, D. M. (2010). When less is more: TPJ and default network deactivation during encoding predicts working memory performance. *NeuroImage, 49*(3), 2638–2648; Li, C. S. R., Yan, P., Bergquist, K. L., & Sinha, R. (2007). Greater activation of the "default" brain regions predicts stop signal errors. *NeuroImage, 38*(3), 640–648.

27 **A group of children with Asperger's** Hayashi, M., Kato, M., Igarashi, K., & Kashima, H. (2008). Superior fluid intelligence in children with Asperger's disorder. *Brain and Cognition, 66*(3), 306–310.

28 **The human brain weighs in at about 1,300 grams** Roth, G., & Dicke, U. (2005). Evolution of the brain and intelligence. *Trends in Cognitive Sciences, 9*(5), 250–257.

29 **newer parts of the brain, like the prefrontal cortex** Schoenemann, P. T. (2006). Evolution of the size and functional areas of the human brain. *Annual Review of Anthropology, 35,* 379–406.

30 **In adult humans, the brain makes up approximately** Aiello, L. C., Bates, N., & Joffe, T. (2001). In defense of the expensive tissue hypothesis. *Evolutionary Anatomy of the Primate Cerebral Cortex.* Cambridge: Cambridge University Press, pp. 57–78; Leonard, W. R., & Robertson, M. L. (1992). Nutritional requirements and human evolution: A bioenergetics model. *American Journal of Human Biology, 4*(2), 179–195.

32 **evolutionary anthropologist Robin Dunbar made the provocative claim** Dunbar, R. I. M. (1998). The social brain hypothesis. *Evolutionary Anthropology, 6,* 178–190.

32 *Neocortex ratio* **refers to the size of Neocortex** literally means "new cortex"; it's the part of the cortex that is most different in structure in primates compared with other mammals.

32 **When the relative size of the neocortex is correlated** Dunbar, R. I. (1992). Neocortex size as a constraint on group size in primates. *Journal of Human Evolution, 22*(6), 469–493; Sawaguchi, T. (1988). Correlations of cerebral indices for "extra" cortical parts and ecological variables in primates. *Brain, Behavior and Evolution, 32*(3), 129–140.

32 **Later work demonstrated that these effects** Schoenemann, P. T. (2006). Evolution of the size and functional areas of the human brain. *Annual Review of Anthropology, 35,* 379–406.

32 **This is referred to as "Dunbar's number"** Dunbar, R. I. (2008). Why humans aren't just Great Apes. *Issues in Ethnology and Anthropology, 3,* 15–33.

32 **village size, estimated from as long ago as 6000 BC** Dunbar, R. I. (1993). Coevolution of neocortical size, group size and language in humans. *Behavioral and Brain Sciences, 16*(4), 681–693.

33 **The most obvious advantage to larger groups** Hill, R. A., & Dunbar, R. I. M. (1998). An evaluation of the roles of predation rate and predation risk as selective pressures on primate grouping behaviour. *Behaviour,* 411–430.

33 **Primates with strong social skills can limit** Silk, J. B. (2002). Using the "F"-word in primatology. *Behaviour,* 421–446.

34 **When we reach Dunbar's number** The formula for determining the number of dyadic permutations is $[N * (N - 1)] / 2$.

Chapter 3: Broken Hearts and Broken Legs

39   **41 percent of respondents indicated that they feared** Bruskin Associates (1973). What are Americans afraid of? *The Bruskin Report,* 53, p. 27.

40   **one of your most painful experiences involved** Jaremka, L. M., Gabriel, S., & Carvallo, M. (2011). What makes us feel the best also makes us feel the worst: The emotional impact of independent and interdependent experiences. *Self and Identity, 10*(1), 44–63.

41   **Giving birth to a baby with a big brain is not easy** Gould, S. J. (1977). *Ontogeny and Phylogeny.* Cambridge, MA: Harvard University Press, Belknap Press; Begun, D., & Walker, A. (1993). The endocast. *The Nariokotome Homo Erectus Skeleton.* Cambridge, MA: Harvard University Press, pp. 326–358; Flinn, M. V., Geary, D. C., & Ward, C. V. (2005). Ecological dominance, social competition, and coalitionary arms races: Why humans evolved extraordinary intelligence. *Evolution and Human Behavior, 26*(1), 10–46; Montagu, A. (1961). Neonatal and infant immaturity in man. *JAMA: Journal of the American Medical Association, 178*(1), 56–57.

41   **The downside to an immature brain** Leigh, S. R., & Park, P. B. (1998). Evolution of human growth prolongation. *American Journal of Physical Anthropology, 107*(3), 331–350.

41   **it is true that the human prefrontal cortex** Gogtay, N., Giedd, J. N., Lusk, L., Hayashi, K. M., Greenstein, D., Vaituzis, A. C., . . . , & Thompson, P. M. (2004). Dynamic mapping of human cortical development during childhood through early adulthood. *Proceedings of the National Academy of Sciences of the United States of America, 101*(21), 8174–8179.

41   **In 1943, Abraham Maslow, a famous New England psychologist** Maslow, A. H. (1943). A theory of human motivation. *Psychological Review, 50*(4), 370.

43   **Without social support, infants will never survive** Baumeister, R. F., & Leary, M. R. (1995). The need to belong: Desire for interpersonal attachments as a fundamental human motivation. *Psychological Bulletin, 117*(3), 497.

44   **pain was responsible for over $60 billion** Stewart, W. F., Ricci, J. A., Chee, E., Morganstein, D., & Lipton, R. (2003). Lost productive time and cost due to common pain conditions in the US workforce. *JAMA: Journal of the American Medical Association, 290*(18), 2443–2454.

44   **Children born with congenital insensitivity to pain** "The Girl Who Can't Feel Pain": http://abcnews.go.com/GMA/OnCall/story?id=1386322.

44   **which animals are able to feel pain** Nordgreen, J., Garner, J. P., Janczak, A. M., Ranheim, B., Muir, W. M., & Horsberg, T. E. (2009). Thermonociception in fish: Effects of two different doses of morphine on thermal threshold and post-test behaviour in goldfish (*Carassius auratus*).

Applied Animal Behaviour Science, 119(1), 101–107; Yue Cottee, S. (2012). Are fish the victims of "speciesism"? A discussion about fear, pain and animal consciousness. *Fish Physiology and Biochemistry,* 1–11.

45   "**A sense of separation is a condition** MacLean, P. D. (1993). Introduction: Perspectives on cingulate cortex in the limbic system. *Neurobiology of Cingulate Cortex and Limbic Thalamus: A Comprehensive Handbook.* Boston: Birkhäuser, pp. 1–19.

45   **I have been studying social pain with my wife** Eisenberger, N. I., & Cole, S. W. (2012). Social neuroscience and health: Neuropsychological mechanisms linking social ties with physical health. *Nature Neuroscience, 15,* 669–674.

45   **pain can be dramatically modified through the power** Crasilneck, H. B., McCranie, E. J., & Jenkins, M. T. (1956). Special indications for hypnosis as a method of anesthesia. *Journal of the American Medical Association, 162*(18), 1606–1608; "Hypnosis, No Anesthetic, for Man's Surgery": http://www.cbsnews.com/2100-500165_162-4033962.html.

45   **In pain experiments, simply expecting that a shock** Sawamoto, N., Honda, M., Okada, T., Hanakawa, T., Kanda, M., Fukuyama, H., . . . , & Shibasaki, H. (2000). Expectation of pain enhances responses to nonpainful somatosensory stimulation in the anterior cingulate cortex and parietal operculum/posterior insula: An event-related functional magnetic resonance imaging study. *Journal of Neuroscience, 20*(19), 7438–7445.

46   **a drink that selectively depletes the brain's serotonin will** Crockett, M. J., Clark, L., Tabibnia, G., Lieberman, M. D., & Robbins, T. W. (2008). Serotonin modulates behavioral reactions to unfairness. *Science, 320,* 1739.

46   **No one has ever broken his arm and confused that** Chen, Z., Williams, K. D., Fitness, J., & Newton, N. C. (2008). When hurt will not heal: Exploring the capacity to relive social and physical pain. *Psychological Science, 19*(8), 789–795.

47   **Psychologists are discovering that language** Zhong, C., Strejcek, B., & Sivanathan, N. (2010). A clean self can render harsh moral judgment. *Journal of Experimental Social Psychology, 46*(5), 859–862.

47   **the language of physical pain is the metaphor du jour** MacDonald, G., & Leary, M. R. (2005). Why does social exclusion hurt? The relationship between social and physical pain. *Psychological Bulletin, 131*(2), 202.

47   **psychologist John Bowlby developed the concept of attachment** Bowlby, J. (1969). *Attachment and loss, volume i: Attachment.* New York: Basic Books.

48   **We all inherited an attachment system** Baumeister, R. F., & Leary, M. R. (1995). The need to belong: Desire for interpersonal attachments as a fundamental human motivation. *Psychological Bulletin, 117*(3), 497.

48   **psychologist Harry Harlow, examined primate attachment** Harlow, H. F. (1958). The nature of love. *American Psychologist, 13,* 673–685.

49   **scientists have discovered *separation distress vocalizations*** Hofer, M. A., & Shair, H. (2004). Ultrasonic vocalization during social interaction and isolation in 2-week-old rats. *Developmental Psychobiology, 11*(5), 495–504; Hennessy, M. B., Nigh, C. K., Sims, M. L., & Long, S. J. (1995). Plasma cortisol and vocalization responses of postweaning age guinea pigs to maternal and sibling separation: Evidence for filial attachment after weaning. *Developmental Psychobiology, 28*(2), 103–115; Boissy, A., & Le Neindre, P. (1997). Behavioral, cardiac and cortisol responses to brief peer separation and reunion in cattle. *Physiology & Behavior, 61*(5), 693–699; Romeyer, A., & Bouissou, M. F. (1992). Assessment of fear reactions in domestic sheep, and influence of breed and rearing conditions. *Applied Animal Behaviour Science, 34*(1), 93–119; Noirot, E. (2004). Ultrasounds and maternal behavior in small rodents. *Developmental Psychobiology, 5*(4), 371–387.

49   **increased production of *cortisol* (a stress hormone)** Coe, C. L., Mendoza, S. P., Smotherman, W. P., & Levine, S. (1978). Mother-infant attachment in the squirrel monkey: Adrenal response to separation. *Behavioral Biology, 22*(2), 256–263; Gamallo, A., Villanua, A., Trancho, G., & Fraile, A. (1986). Stress adaptation and adrenal activity in isolated and crowded rats. *Physiology & Behavior, 36*(2), 217–221; Parrott, R. F., Houpt, K. A., & Misson, B. H. (1988). Modification of the responses of sheep to isolation stress by the use of mirror panels. *Applied Animal Behaviour Science, 19*(3), 331–338.

49   **Children under the age of five** Douglas, W. B. (1975). Early hospital admissions and later disturbances of behaviour and learning. *Developmental Medicine & Child Neurology, 17*(4), 456–480.

49   **children who lose a parent** Luecken, L. J. (1998). Childhood attachment and loss experiences affect adult cardiovascular and cortisol function. *Psychosomatic Medicine, 60*(6), 765–772.

49   **This type of early childhood stressor** Hanson, J. L., Chung, M. K., Avants, B. B., Shirtcliff, E. A., Gee, J. C., Davidson, R. J., & Pollak, S. D. (2010). Early stress is associated with alterations in the orbitofrontal cortex: A tensor-based morphometry investigation of brain structure and behavioral risk. *Journal of Neuroscience, 30*(22), 7466–7472.

49   **Opioids are the brain's natural painkillers** Panksepp, J., Herman, B. H., Conner, R., Bishop, P., & Scott, J. P. (1978). The biology of social attachments: Opiates alleviate separation distress. *Biological Psychiatry, 13,* 607–613.

50   **nonsedating levels of opiates** Carden, S. E., & Hofer, M. A. (1990). Independence of benzodiazepine and opiate action in the suppression of isolation distress in rat pups. *Behavioral Neuroscience, 104*(1), 160–166;

Herman, B. H., & Panksepp, J. (1978). Effects of morphine and naloxone on separation distress and approach attachment: Evidence for opiate mediation of social affect. *Pharmacology Biochemistry and Behavior, 9*(2), 213–220; Kalin, N. H., Shelton, S. E., & Barksdale, C. M. (1988). Opiate modulation of separation-induced distress in non-human primates. *Brain Research, 440*(2), 285–292.

50 **reconnection between mother and infant** Kalin, N. H., Shelton, S. E., & Lynn, D. E. (1995). Opiate systems in mother and infant primates coordinate intimate contact during reunion. *Psychoneuroendocrinology, 20*(7), 735–742; Keverne, E. B., Martensz, N. D., & Tuite, B. (1989). Beta-endorphin concentrations in cerebrospinal fluid of monkeys are influenced by grooming relationships. *Psychoneuroendocrinology, 14*(1), 155–161.

50 **we do not conduct experiments with humans** It has been suggested that opiate addiction sometimes reflects an attempt to replace the natural opioids released by the body as a result of social and physical contact [MacLean, P. D. (1985). Brain evolution relating to family, play, and the separation call. *Archives of General Psychiatry, 42*(4), 405.]).

51 **There are four reasons why an investigation** MacLean, P. D. (1985). Brain evolution relating to family, play, and the separation call. *Archives of General Psychiatry, 42*(4), 405.

51 **the ACC has the highest density of opioid receptors** Wise, S. P., & Herkenham, M. (1982). Opiate receptor distribution in the cerebral cortex of the Rhesus monkey. *Science, 218*(4570), 387.

52 **The *sensory aspects of pain* tell us where** Talbot, J. D., Marrett, S., Evans, A. C., & Meyer, E. (1991). Multiple representations of pain in human cerebral cortex. *Science, 251*(4999), 1355–1358; Rainville, P., Duncan, G. H., Price, D. D., Carrier, B., & Bushnell, M. C. (1997). Pain affect encoded in human anterior cingulate but not somatosensory cortex. *Science, 277*(5328), 968–971.

52 **many different brain regions are working together** Kosslyn, S. M. (1992). *Wet Mind*. New York: Free Press.

53 **there are patients with damage to motion perception centers** Zihl, J., Von Cramon, D., & Mai, N. (1983). Selective disturbance of movement vision after bilateral brain damage. *Brain, 106*(2), 313–340.

53 **This surgery has been successfully used** Whitty, C. W., Duffield, J. E., & Cairns, H. (1952). Anterior cingulectomy in the treatment of mental disease. *Lancet, 1*(6706), 475; Le Beau, J. (1954). Anterior cingulectomy in man. *Journal of Neurosurgery, 11*(3), 268; Whitty, C. W. M. (1955). Effects of anterior cingulectomy in man. *Proceedings of the Royal Society of Medicine, 48*(6), 463; Steele, J. D., Christmas, D., Eljamel, M. S., & Matthews, K. (2008). Anterior cingulotomy for major depression:

Clinical outcome and relationship to lesion characteristics. *Biological Psychiatry, 63*(7), 670–677.

53 **They report that they still feel pain** Foltz, E. L., & White Jr., L. E. (1962). Pain "relief" by frontal cingulumotomy. *Journal of Neurosurgery, 19,* 89.

53 **As painful stimulation was applied to his left arm** Ploner, M., Freund, H. J., & Schnitzler, A. (1999). Pain affect without pain sensation in a patient with a postcentral lesion. *Pain, 81*(1), 211–214.

54 **Neuroscientist Paul MacLean experimented with the effects of *lesioning*** MacLean, P. D., & Newman, J. D. (1988). Role of midline frontolimbic cortex in production of the isolation call of squirrel monkeys. *Brain Research, 450*(1), 111–123.

55 **When the dACC is stimulated in rhesus monkeys** Robinson, B. W. (1967). Vocalization evoked from forebrain in *Macaca mulatta*. *Physiology & Behavior, 2*(4), 345–354; Smith, W. K. (1945). The functional significance of the rostral cingular cortex as revealed by its responses to electrical excitation. *Journal of Neurophysiology, 8,* 241–254.

55 **female rats were treated in one of three ways** Stamm, J. S. (1955). The function of the median cerebral cortex in maternal behavior of rats. *Journal of Comparative and Physiological Psychology, 48*(4), 347; see also Murphy, M. R., MacLean, P. D., & Hamilton, S. C. (1981). Species-typical behavior of hamsters deprived from birth of the neocortex. *Science, 213,* 459–461.

56 **Kip Williams's paradigm was called *Cyberball*** Williams, K. D., Cheung, C. K., & Choi, W. (2000). Cyberostracism: Effects of being ignored over the Internet. *Journal of Personality and Social Psychology, 79*(5), 748; Williams, K. D. (2007). Ostracism. *Annual Review of Psychology, 58,* 425–452

57 **We had people play *Cyberball*** Eisenberger, N. I., Lieberman, M. D., & Williams, K. D. (2003). Does rejection hurt? An fMRI study of social exclusion. *Science, 302,* 290–292.

58 **there was a single moment at which point we knew** Lieberman, M. D., Jarcho, J. M., Berman, S., Naliboff, B., Suyenobu, B. Y., Mandelkern, M., & Mayer, E. (2004). The neural correlates of placebo effects: A disruption account. *NeuroImage, 22,* 447–455.

58 **in the social pain study, participants who activated** See Chapter 9 for more on the role of the right ventrolateral prefrontal cortex in self-control and emotion regulation.

59 **Looking at the screens, side by side** Eisenberger, N. I., & Lieberman, M. D. (2004). Why it hurts to be left out: The neurocognitive overlap between physical and social pain. *Trends in Cognitive Sciences, 8,* 294–300.

59   **lots of scientists didn't buy our findings** Eisenberger, N. I., & Cole, S. W. (2012). Social neuroscience and health: Neuropsychological mechanisms linking social ties with physical health. *Nature Neuroscience, 15,* 669–674.

59   **In the mid- to late 1990s, several neuroimaging studies were published** Botvinick, M., Nystrom, L. E., Fissell, K., Carter, C. S., & Cohen, J. D. (1999). Conflict monitoring versus selection-for-action in anterior cingulate cortex. *Nature, 402*(6758), 179–181; Carter, C. S., Braver, T. S., Barch, D. M., Botvinick, M. M., Noll, D., & Cohen, J. D. (1998). Anterior cingulate cortex, error detection, and the online monitoring of performance. *Science, 280*(5364), 747–749.

60   **a seminal paper on the function of the dACC** Bush, G., Luu, P., & Posner, M. I. (2000). Cognitive and emotional influences in anterior cingulate cortex. *Trends in Cognitive Sciences, 4*(6), 215–222.

60   **Psychologists have long enjoyed dichotomizing processes** Tetlock, P. E., & Levi, A. (1982). Attribution bias: On the inconclusiveness of the cognition-motivation debate. *Journal of Experimental Social Psychology, 18*(1), 68–88.

60   **several other neuroimaging papers of emotion or pain distress** Morris, J. S., Frith, C. D., Perrett, D. I., Rowland, D., Young, A. W., Calder, A. J., & Dolan, R. J. (1996). A differential neural response in the human amygdala to fearful and happy facial expressions. *Nature, 383,* 812–815; Morris, J. S., Friston, K. J., Büchel, C., Frith, C. D., Young, A. W., Calder, A. J., & Dolan, R. J. (1998). A neuromodulatory role for the human amygdala in processing emotional facial expressions. *Brain, 121*(1), 47–57; Kimbrell, T. A., George, M. S., Parekh, P. I., Ketter, T. A., Podell, D. M., Danielson, A. L., . . . , & Post, R. M. (1999). Regional brain activity during transient self-induced anxiety and anger in healthy adults. *Biological Psychiatry, 46*(4), 454–465; Lane, R. D., Reiman, E. M., Axelrod, B., Yun, L. S., Holmes, A., & Schwartz, G. E. (1998). Neural correlates of levels of emotional awareness: Evidence of an interaction between emotion and attention in the anterior cingulate cortex. *Journal of Cognitive Neuroscience, 10*(4), 525–535; Schneider, F., Grodd, W., Weiss, U., Klose, U., Mayer, K. R., Nägele, T., & Gur, R. C. (1997). Functional MRI reveals left amygdala activation during emotion. *Psychiatry Research: Neuroimaging, 76*(2–3), 75–82; Teasdale, J. D., Howard, R. J., Cox, S. G., Ha, Y., Brammer, M. J., Williams, S. C., & Checkley, S. A. (1999). Functional MRI study of the cognitive generation of affect. *American Journal of Psychiatry, 156*(2), 209–215; Sawamoto, N., Honda, M., Okada, T., Hanakawa, T., Kanda, M., Fukuyama, H., . . . , & Shibasaki, H. (2000). Expectation of pain enhances responses to nonpainful somatosensory stimulation in the anterior cingulate cortex and parietal operculum/posterior insula: An

event-related functional magnetic resonance imaging study. *Journal of Neuroscience, 20*(19), 7438–7445; Talbot, J. D., Marrett, S., Evans, A. C., & Meyer, E. (1991). Multiple representations of pain in human cerebral cortex. *Science, 251,* 1355–1358; Jones, A. K. P., Brown, W. D., Friston, K. J., Qi, L. Y., & Frackowiak, R. S. J. (1991). Cortical and subcortical localization of response to pain in man using positron emission tomography. *Proceedings of the Royal Society of London. Series B: Biological Sciences, 244*(1309), 39–44; Coghill, R. C., Talbot, J. D., Evans, A. C., Meyer, E., Gjedde, A., Bushnell, M. C., & Duncan, G. H. (1994). Distributed processing of pain and vibration by the human brain. *Journal of Neuroscience, 14*(7), 4095–4108; Casey, K. L., Minoshima, S., Berger, K. L., Koeppe, R. A., Morrow, T. J., & Frey, K. A. (1994). Positron emission tomographic analysis of cerebral structures activated specifically by repetitive noxious heat stimuli. *Journal of Neurophysiology, 71*(2), 802–807; Rainville, P., Duncan, G. H., Price, D. D., Carrier, B., & Bushnell, M. C. (1997). Pain affect encoded in human anterior cingulate but not somatosensory cortex. *Science, 277*(5328), 968–971.

61 **a paper on a new model of dACC function** Eisenberger, N. I., & Lieberman, M. D. (2004). Why it hurts to be left out: The neurocognitive overlap between physical and social pain. *Trends in Cognitive Sciences, 8,* 294–300. For more recent reviews taking a similar view, see Shackman, A. J., Salomons, T. V., Slagter, H. A., Fox, A. S., Winter, J. J., & Davidson, R. J. (2011). The integration of negative affect, pain and cognitive control in the cingulate cortex. *Nature Reviews Neuroscience, 12*(3), 154–167; Etkin, A., Egner, T., & Kalisch, R. (2011). Emotional processing in anterior cingulate and medial prefrontal cortex. *Trends in Cognitive Sciences, 15*(2), 85–93.

61 **everyone figures out to use the metal knocker** Gilbert, D. T., Lieberman, M. D., Morewedge, C. K., & Wilson, T. D. (2004). The peculiar longevity of things not so bad. *Psychological Science, 15,* 14–19.

62 **a conflict/error detection procedure called the *stop-signal task*** Spunt, R. P., Lieberman, M. D., Cohen, J. R., & Eisenberger, N. I. (2012). The phenomenology of error processing: The dorsal anterior cingulate response to stop-signal errors tracks reports of negative affect. *Journal of Cognitive Neuroscience, 24,* 1753–1765; see also Botvinick, M. M. (2007). Conflict monitoring and decision making: Reconciling two perspectives on anterior cingulate function. *Cognitive, Affective, & Behavioral Neuroscience, 7,* 356–366.

64 **Our basic findings linking social exclusion** Masten, C. L., Telzer, E. H., Fuligni, A. J., Lieberman, M. D., & Eisenberger, N. I. (2012). Time spent with friends in adolescence relates to less neural sensitivity to later peer rejection. *Social Cognitive and Affective Neuroscience, 7*(1), 106–114; Bolling, D. Z., Pitskel, N. B., Deen, B., Crowley, M. J., McPartland, J. C.,

Mayes, L. C., & Pelphrey, K. A. (2011). Dissociable brain mechanisms for processing social exclusion and rule violation. *NeuroImage, 54*(3), 2462–2471; Krill, A., & Platek, S. M. (2009). In-group and out-group membership mediates anterior cingulate activation to social exclusion. *Frontiers in Evolutionary Neuroscience, 1,* 1–7; Bolling, D. Z., Pelphrey, K. A., & Vander Wyk, B. C. (2012). Differential brain responses to social exclusion by one's own versus opposite-gender peers. *Social Neuroscience, 7*(4), 331–346; Wager, T. D., van Ast, V. A., Hughes, B. L., Davidson, M. L., Lindquist, M. A., & Ochsner, K. N. (2009). Brain mediators of cardiovascular responses to social threat, part II: Prefrontal-subcortical pathways and relationship with anxiety. *NeuroImage, 47*(3), 836–851; Burklund, L. J., Eisenberger, N. I., & Lieberman, M. D. (2007). The face of rejection: Rejection sensitivity moderates dorsal anterior cingulate activity to disapproving facial expressions. *Social Neuroscience, 2*(3-4), 238–253; Fisher, H. E., Brown, L. L., Aron, A., Strong, G., & Mashek, D. (2010). Reward, addiction, and emotion regulation systems associated with rejection in love. *Journal of Neurophysiology, 104*(1), 51–60; Kross, E., Berman, M. G., Mischel, W., Smith, E. E., & Wager, T. D. (2011). Social rejection shares somatosensory representations with physical pain. *Proceedings of the National Academy of Sciences, 108*(15), 6270–6275; O'Connor, M. F., Wellisch, D. K., Stanton, A. L., Eisenberger, N. I., Irwin, M. R., & Lieberman, M. D. (2008). Craving love? Enduring grief activates brain's reward center. *NeuroImage, 42*(2), 969–972; Gündel, H., O'Connor, M. F., Littrell, L., Fort, C., & Lane, R. D. (2003). Functional neuroanatomy of grief: An fMRI study. *American Journal of Psychiatry, 160*(11), 1946–1953; Kersting, A., Ohrmann, P., Pedersen, A., Kroker, K., Samberg, D., Bauer, J., . . . , & Suslow, T. (2009). Neural activation underlying acute grief in women after the loss of an unborn child. *American Journal of Psychiatry, 166*(12), 1402–1410; Onoda, K., Okamoto, Y., Nakashima, K. I., Nittono, H., Yoshimura, S., Yamawaki, S., . . . , & Ura, M. (2010). Does low self-esteem enhance social pain? The relationship between trait self-esteem and anterior cingulate cortex activation induced by ostracism. *Social Cognitive and Affective Neuroscience, 5*(4), 385–391; Eisenberger, N. I., Inagaki, T. K., Muscatell, K. A., Byrne Haltom, K. E., & Leary, M. R. (2011). The neural sociometer: Brain mechanisms underlying state self-esteem. *Journal of Cognitive Neuroscience, 23*(11), 3448–3455.

64 **over-the-counter painkillers would reduce social pain** DeWall, C. N., MacDonald, G., Webster, G. D., Masten, C. L., Baumeister, R. F., Powell, C., Combs, D., Schurtz, D. R., Stillman, T. F., Tice, D. M., & Eisenberger, N. I. (2010). Acetaminophen reduces social pain: Behavioral and neural evidence. *Psychological Science, 21,* 931–937.

65   Mice that have been bred to lack the mu-opioid receptor Sora I., et al. (1997). Opiate receptor knockout mice define mu receptor roles in endogenous nociceptive responses and morphine-induced analgesia. *Proceedings of the National Academy of Sciences of the United States of America, 94,* 1544–1549.

65   We inherit one allele from our mother Sia, A. T., et al. (2008). A118G single nucleotide polymorphism of human mu-opioid receptor gene influences pain perception and patient-controlled intravenous morphine consumption after intrathecal morphine for postcesarean analgesia. *Anesthesiology, 109,* 520–526; Coulbault, L., et al. (2006). Environmental and genetic factors associated with morphine response in the postoperative period. *Clinical Pharmacology & Therapeutics, 79,* 316–324; Chou, W. Y., et al. (2006). Association of mu-opioid receptor gene polymorphism (A118G) with variations in morphine consumption for analgesia after total knee arthroplasty. *Acta Anaesthesiology Scandinivaca, 50,* 787–792.

66   Genetic samples were obtained from a group of individuals Way, B. M., Taylor, S. E., & Eisenberger, N. I. (2009). Variation in the mu-opioid receptor gene (OPRM1) is associated with dispositional and neural sensitivity to social rejection. *Proceedings of the National Academy of Sciences of the United Stated of America, 106,* 15079–15084.

67   The human visual system makes various assumptions James, W. (1890/1950). *The Principles of Psychology.* New York: Dover.

68   Kip Williams found that even when he told people Zadro, L., Williams, K. D., & Richardson, R. (2004). How low can you go? Ostracism by a computer is sufficient to lower self-reported levels of belonging, control, self-esteem, and meaningful existence. *Journal of Experimental Social Psychology, 40*(4), 560–567.

69   why would all those others watch the bully Kaltiala-Heino, R., Rimpelä, M., Marttunen, M., Rimpelä, A., & Rantanen, P. (1999). Bullying, depression, and suicidal ideation in Finnish adolescents: School survey. *Bmj, 319*(7206), 348–351; Juvonen, J., & Galván, A. (2009). Bullying as a means to foster compliance. In M. Harris (Ed.). *Bullying, Rejection and Peer Victimization: A Social Cognitive Neuroscience Perspective.* New York: Springer, pp. 299–318.

69   about 10 percent of students are bullied Fleming, L. C., & Jacobsen, K. H. (2009). Bullying and symptoms of depression in Chilean middle school students. *Journal of School Health, 79*(3), 130–137; Wolke, D., Woods, S., Stanford, K., & Schulz, H. (2001). Bullying and victimization of primary school children in England and Germany: Prevalence and school factors. *British Journal of Psychology, 92*(4), 673–696; Kaltiala-Heino, R., Rimpelä, M., Marttunen, M., Rimpelä, A., & Rantanen, P.

(1999). Bullying, depression, and suicidal ideation in Finnish adolescents: School survey. *Bmj, 319*(7206), 348–351; Kim, Y. S., Koh, Y. J., & Leventhal, B. (2005). School bullying and suicidal risk in Korean middle school students. *Pediatrics, 115*(2), 357–363.

69   **they involve belittling comments** Nansel, T. R., Overpeck, M., Pilla, R. S., Ruan, W. J., Simons-Morton, B., & Scheidt, P. (2001). Bullying behaviors among US youth. *JAMA: Journal of the American Medical Association, 285*(16), 2094–2100.

69   **They think about committing suicide more** Klomek, A. B., Marrocco, F., Kleinman, M., Schonfeld, I. S., & Gould, M. S. (2007). Bullying, depression, and suicidality in adolescents. *Journal of the American Academy of Child & Adolescent Psychiatry, 46*(1), 40.

69   **A 1989 Finnish study assessed the level of victimization** Klomek, A. B., Sourander, A., Niemelä, S., Kumpulainen, K., Piha, J., Tamminen, T., . . . , & Gould, M. S. (2009). Childhood bullying behaviors as a risk for suicide attempts and completed suicides: A population-based birth cohort study. *Journal of the American Academy of Child & Adolescent Psychiatry, 48*(3), 254–261.

69   **Suicide-related thoughts are actually quite similar** Smith, M. T., Edwards, R. R., Robinson, R. C., & Dworkin, R. H. (2004). Suicidal ideation, plans, and attempts in chronic pain patients: Factors associated with increased risk. *Pain, 111,* 201–208.

Chapter 4: Fairness Tastes like Chocolate

72   **and independently performed an anagram** Hegtvedt, K. A., & Killian, C. (1999). Fairness and emotions: Reactions to the process and outcomes of negotiations. *Social Forces, 78*(1), 269–302.

73   **Psychologist Tom Tyler found that defendants in court cases** Tyler, T. R. (1984). The role of perceived injustice in defendants' evaluations of their courtroom experience. *Law & Society Review, 18,* 51.

73   **evidence for or against the notion that fairness** Tabibnia, G., Satpute, A. B., & Lieberman, M. D. (2008). The sunny side of fairness: Preference for fairness activates reward circuitry (and disregarding unfairness activates self-control circuitry). *Psychological Science, 19,* 339–347.

74   **studies typically observe activity in the anterior insula and the dACC** Sanfey, A. G., Rilling, J. K., Aronson, J. A., Nystrom, L. E., & Cohen, J. D. (2003). The neural basis of economic decision-making in the ultimatum game. *Science, 300*(5626), 1755–1758; Civai, C., Crescentini, C., Rustichini, A., & Rumiati, R. I. (2012). Equality versus self-interest in the brain: Differential roles of anterior insula and medial prefrontal cortex. *NeuroImage, 62,* 102–112.

74 a group of researchers from Cal Tech examined the neural responses
Tricomi, E., Rangel, A., Camerer, C. F., & O'Doherty, J. P. (2010).
Neural evidence for inequality-averse social preferences. *Nature*,
463(7284), 1089–1091.

75 these are referred to as *social rewards* Lieberman, M. D., & Eisenberger,
N. I. (2009). Pains and pleasures of social life. *Science, 323*, 890–891.

76 Signs that others like, admire, and love us Baumeister, R. F., & Leary,
M. R. (1995). The need to belong: Desire for interpersonal attachments
as a fundamental human motivation. *Psychological Bulletin, 117*(3), 497.

76 asked participants for permission to contact their friends Inagaki, T. K.,
& Eisenberger, N. I. (in press). Shared neural mechanisms underlying
social warmth and physical warmth, *Psychological Science*.

77 looked at how rewarding these touching statements really were Castle, E.,
& Lieberman, M. D. (unpublished data). How much would you pay to
hear "I love you"?

77 our reactions to getting this rarely shared positive feedback Guyer, A. E.,
Choate, V. R., Pine, D. S., & Nelson, E. E. (2012). Neural circuitry
underlying affective response to peer feedback in adolescence. *Social
Cognitive and Affective Neuroscience, 7*(1), 81–92; Davey, C. G., Allen,
N. B., Harrison, B. J., Dwyer, D. B., & Yücel, M. (2010). Being liked
activates primary reward and midline self-related brain regions. *Human
Brain Mapping, 31*(4), 660–668.

77 participants in the scanner saw that strangers Izuma, K., Saito, D. N.,
& Sadato, N. (2008). Processing of social and monetary rewards in the
human striatum. *Neuron, 58*(2), 284.

78 praise taps into the same reinforcement system Baumeister, R. F.,
Campbell, J. D., Krueger, J. I., & Vohs, K. D. (2003). Does high self-
esteem cause better performance, interpersonal success, happiness, or
healthier lifestyles? *Psychological Science in the Public Interest, 4*(1), 1–44.

79 Rewards can be divided Hull, C. L. (1952). *A Behavior System: An
Introduction to Behavior Theory Concerning the Individual Organism*.
New Haven: Yale University Press.

80 The red patch is not intrinsically rewarding Schultz, W., Dayan, P., &
Montague, P. R. (1997). A neural substrate of prediction and reward.
Science, 275(5306), 1593–1599.

81 humans are supercooperators Melis, A. P., Semmann, D., Melis, A. P., &
Semmann, D. (2010). How is human cooperation different? *Philosophical
Transactions of the Royal Society B: Biological Sciences, 365*(1553), 2663–
2674; Nowak, M., & Highfield, R. (2012). *SuperCooperators: Altruism,
Evolution, and Why We Need Each Other to Succeed*. New York: Free Press.

81 The *principle of reciprocity* is one of the strongest Cialdini, R. B. (2001).
Influence: Science and Practice (Vol. 4). Boston: Allyn & Bacon; Burger,

J. M., Sanchez, J., Imberi, J. E., & Grande, L. R. (2009). The norm of reciprocity as an internalized social norm: Returning favors even when no one finds out. *Social Influence, 4*(1), 11–17.

81 **By performing a small favor for you** Regan, R. T. (1971). Effects of a favor and liking on compliance. *Journal of Experimental Social Psychology, 7,* 627–639.

82 **a game called the *Prisoner's Dilemma*** I highly recommend doing a YouTube search for "golden balls," a British game show based on the Prisoner's Dilemma. The top few hits are highly entertaining.

83 **people still choose to cooperate** Hayashi, N., Ostrom, E., Walker, J., & Yamagishi, T. (1999). Reciprocity, trust, and the sense of control: A cross-societal study. *Rationality and Society, 11*(1), 27–46; Kiyonari, T., Tanida, S., & Yamagishi, T. (2000). Social exchange and reciprocity: Confusion or a heuristic? *Evolution and Human Behavior, 21*(6), 411–427.

83 **How can we explain why folks cooperate** Kiyonari, T., Tanida, S., & Yamagishi, T. (2000). Social exchange and reciprocity: Confusion or a heuristic? *Evolution and Human Behavior, 21*(6), 411–427.

83 **"the first principle of economics** Edgeworth, F. Y. (1881). *Mathematical Psychics: An Essay on the Application of Mathematics to the Moral Sciences.* London: Kegan Paul, p. 104.

83 **"no other end, in all his actions** Hume (1898/1754, p.117). Hume, D. (2001/1754). *An Enquiry Concerning Human Understanding* (Vol. 3). New York: Oxford University Press, p. 117.

83 **"every man is presumed to seek** Hobbes, T. (1969/1651). *Leviathan* (part iii). Aldershot, England: Scolar Press.

83 **known as the *axiom of self-interest*** Hollander, S. (1977). Adam Smith and the self-interest axiom. *Journal of Law and Economics, 20*(1), 133–152.

84 **What is surprising, though, is that** Hayashi, N., Ostrom, E., Walker, J., & Yamagishi, T. (1999). Reciprocity, trust, and the sense of control: A cross-societal study. *Rationality and Society, 11*(1), 27–46.

84 **in addition to being self-interested** Fehr, E., & Camerer, C. F. (2007). Social neuroeconomics: The neural circuitry of social preferences. *Trends in Cognitive Sciences, 11*(10), 419–427.

84 **people made decisions counter to their own self-interest** Henrich, J., Boyd, R., Bowles, S., Camerer, C., Fehr, E., Gintis, H., . . . , & Tracer, D. (2005). "Economic man" in cross-cultural perspective: Behavioral experiments in 15 small-scale societies. *Behavioral and Brain Sciences, 28*(6), 795–814.

85 **"try to teach generosity and altruism** Dawkins, R. (1976). *The Selfish Gene.* Oxford: Oxford University Press.

85 **We know what the brain looks like** Spitzer, M., Fischbacher, U.,

Herrnberger, B., Grön, G., & Fehr, E. (2007). The neural signature of social norm compliance. *Neuron*, *56*(1), 185–196; O'Doherty, J. P., Buchanan, T. W., Seymour, B., & Dolan, R. J. (2006). Predictive neural coding of reward preference involves dissociable responses in human ventral midbrain and ventral striatum. *Neuron*, *49*(1), 157.

85 **in the minds of people as they cooperate or defect** Rilling, J. K., Gutman, D. A., Zeh, T. R., Pagnoni, G., Berns, G. S., & Kilts, C. D. (2002). A neural basis for social cooperation. *Neuron*, *35*(2), 395–405.

86 **ruling out long-term strategies like reputation building** Rilling, J. K., Sanfey, A. G., Aronson, J. A., Nystrom, L. E., & Cohen, J. D. (2004). Opposing BOLD responses to reciprocated and unreciprocated altruism in putative reward pathways. *Neuroreport*, *15*(16), 2539–2243.

86 **In Isaac Asimov's book *The End of the Eternity*** Asimov, I. (2010/1955). *The End of Eternity*. New York: Tor Books, pp. 117–118.

87 **"Scratch an 'altruist' and watch a 'hypocrite' bleed"** Ghiselin, M. T. (1974). *The Economy of Nature and the Evolution of Sex* (Vol. 247). Berkeley: University of California Press; Dawkins, R. (1976). *The Selfish Gene*. Oxford: Oxford University Press.

87 **there may be a hidden selfish motivation** Batson, C. D. (1991). *The Altruism Question: Toward a Social-Psychological Answer*. Hillsdale, NJ: Lawrence Erlbaum Associates, p. 116.

89 **the psychological mechanism that motivates us to selflessly help** Wilson, E. O. (2012). *The Social Conquest of Earth*. New York: Liveright.

89 **what some call the warm glow of altruistic behavior** Andreoni, J. (1990). Impure altruism and donations to public goods: A theory of warm-glow giving. *Economic Journal*, *100*(401), 464–477.

89 **to be selfish, you should do it in a very intelligent way** Lama, D. (1994). *The Way to Freedom*. New York: HarperCollins, p. 154.

89 **an fMRI study looking at the activity in the brain** Moll, J., Krueger, F., Zahn, R., Pardini, M., de Oliveira-Souza, R., & Grafman, J. (2006). Human fronto-mesolimbic networks guide decisions about charitable donation. *Proceedings of the National Academy of Sciences*, *103*(42), 15623–15628; Harbaugh, W. T., Mayr, U., & Burghart, D. R. (2007). Neural responses to taxation and voluntary giving reveal motives for charitable donations. *Science*, *316*(5831), 1622–1625.

90 **the most selfish people on the planet: teenagers** Telzer, E. H., Masten, C. L., Berkman, E. T., Lieberman, M. D., & Fuligni, A. J. (2010). Gaining while giving: An fMRI study of the rewards of family assistance among White and Latino youth. *Social Neuroscience*, *5*, 508–518.

90 **examined supportive behavior between boyfriends and girlfriends** Inagaki, T. K., & Eisenberger, N. I. (2012). Neural correlates of giving support to a loved one. *Psychosomatic Medicine*, *74*, 3–7.

91   **our support of others could contribute significantly to our well-being**
      Brown, S. L., Nesse, R. M., Vinokur, A. D., & Smith, D. M. (2003).
      Providing social support may be more beneficial than receiving it: Results
      from a prospective study of mortality. *Psychological Science, 14*(4), 320–
      327.

91   **Adam Smith, one of the founders of modern economics** Smith, A. (1776).
      An Inquiry into the Nature and Causes of the Wealth of Nations. London:
      W. Strahan and T. Cadell.

91   **"How selfish soever man may be supposed** Smith, A. (1759). *The Theory
      of Moral Sentiments.* Edinburgh: A. Kincaid and J. Bell.

92   **Many mammalian species have shown opioid-linked pleasure responses**
      Keverne, E. B., Martensz, N. D., & Tuite, B. (1989). Beta-endorphin
      concentrations in cerebrospinal fluid of monkeys are influenced by
      grooming relationships. *Psychoneuroendocrinology, 14*(1), 155–161.

92   **in humans most of our grooming is verbal** Dunbar, R. (1998). Theory
      of mind and the evolution of language. In J. Hurford, M. Studdart-
      Kennedy, & C. Knight (Eds.). *Approaches to the Evolution of Language.*
      Cambridge: Cambridge University Press, pp. 92–110.

92   **an incredibly reinforcing signal to receive** Seltzer, L. J., Ziegler, T. E., &
      Pollak, S. D. (2010). Social vocalizations can release oxytocin in humans.
      Proceedings of the Royal Society B: Biological Sciences, 277(1694), 2661–
      2666.

92   **Mammalian mothers of all stripes are jumpstarted into caregiving** Broad,
      K. D., Curley, J. P., & Keverne, E. B. (2006). Mother-infant bonding
      and the evolution of mammalian social relationships. *Philosophical
      Transactions of the Royal Society B: Biological Sciences, 361*(1476), 2199–
      2214.

92   **Oxytocin's primary physiological contribution is to facilitate labor**
      Soloff, M. S., Alexandrova, M., & Fernstrom, M. J. (1979). Oxytocin
      receptors: Triggers for parturition and lactation? *Science, 204*(4399),
      1313.

93   **In contrast, the effects of oxytocin are better characterized as modifying**
      Depue, R. A., & Morrone-Strupinsky, J. V. (2005). A neurobehavioral
      model of affiliative bonding: Implications for conceptualizing a human
      trait of affiliation. *Behavioral and Brain Sciences, 28*(3), 313–349.

94   **oxytocin released in the ventral tegmental area leads to** Febo, M.,
      Numan, M., & Ferris, C. F. (2005). Functional magnetic resonance
      imaging shows oxytocin activates brain regions associated with mother-
      pup bonding during suckling. *Journal of Neuroscience, 25*(50), 11637–
      11644; Shahrokh, D. K., Zhang, T. Y., Diorio, J., Gratton, A., & Meaney,
      M. J. (2010). Oxytocin-dopamine interactions mediate variations in
      maternal behavior in the rat. *Neuroendocrinology, 151*(5), 2276–2286.

94   **Fearlessness appears to be influenced by oxytocin interactions** Leng,

G., Meddle, S. L., & Douglas, A. J. (2008). Oxytocin and the maternal brain. *Current Opinion in Pharmacology, 8*(6), 731–734.

94 **Both oxytocin and the septal region of the brain are involved** Gordon, I., Zagoory-Sharon, O., Schneiderman, I., Leckman, J. F., Weller, A., & Feldman, R. (2008). Oxytocin and cortisol in romantically unattached young adults: Associations with bonding and psychological distress. *Psychophysiology, 45*(3), 349–352; Bartz, J. A., Zaki, J., Bolger, N., & Ochsner, K. N. (2011). Social effects of oxytocin in humans: Context and person matter. *Trends in Cognitive Sciences, 15*(7), 301–309.

94 **Although there are great similarities in how oxytocin promotes care** Numan, M., & Sheehan, T. P. (1997). Neuroanatomical circuitry for mammalian maternal behavior. *Annals of the New York Academy of Sciences, 807*(1), 101–125.

94 **A mother sheep will attack an unrelated baby lamb** Broad, K. D., Curley, J. P., & Keverne, E. B. (2006). Mother-infant bonding and the evolution of mammalian social relationships. *Philosophical Transactions of the Royal Society B: Biological Sciences, 361*(1476), 2199–2214.

94 **Administering oxytocin has been shown to increase generosity** Kosfeld, M., Heinrichs, M., Zak, P. J., Fischbacher, U., & Fehr, E. (2005). Oxytocin increases trust in humans. *Nature, 435*(7042), 673–676; Zak, P. J., Stanton, A. A., & Ahmadi, S. (2007). Oxytocin increases generosity in humans. *PLOS One, 2*(11), e1128.

94 **administering oxytocin leads to more aggressive responses** De Dreu, C. K., Greer, L. L., Van Kleef, G. A., Shalvi, S., & Handgraaf, M. J. (2011). Oxytocin promotes human ethnocentrism. *Proceedings of the National Academy of Sciences, 108*(4), 1262–1266.

95 **Administering oxytocin in humans facilitates caregiving** Kosfeld, M., Heinrichs, M., Zak, P. J., Fischbacher, U., & Fehr, E. (2005). Oxytocin increases trust in humans. *Nature, 435*(7042), 673–676; Fershtman, C., Gneezy, U., & Verboven, F. (2005). Discrimination and nepotism: The efficiency of the anonymity rule. *Journal of Legal Studies, 34*(2), 371–396.

96 **Dale Miller, a social psychologist at Stanford University** Miller, D. T. (1999). The norm of self-interest. *American Psychologist, 54*(12), 1053.

96 **we have learned that people are self-interested** Miller, D. T., & Ratner, R. K. (1998). The disparity between the actual and assumed power of self-interest. *Journal of Personality and Social Psychology, 74*(1), 53.

97 **when people are asked why they have engaged in prosocial behaviors** Wuthnow, R. (1991). *Acts of Compassion: Caring for Others and Helping Ourselves.* Princeton, NJ: Princeton University Press.

97 **in another of Miller's studies** Holmes, J. G., Miller, D. T., & Lerner, M. J. (2002). Committing altruism under the cloak of self-interest: The exchange fiction. *Journal of Experimental Social Psychology, 38*(2), 144–151.

98 "The Americans . . . are fond of explaining almost all the actions de Tocqueville, A. (1958/1835). *Democracy in America.* New York: Vintage.

98 Pain and pleasure are the driving forces of our motivational lives Freud, S. (1950/1920). *Beyond the Pleasure Principle.* New York: Liveright.

99 The severing of a social bond Beck, A. T., Laude, R., & Bohnert, M. (1974). Ideational components of anxiety neurosis. *Archives of General Psychiatry, 31,* 319–325; Brown, G. W., & Harris, T. (2001). *Social Origins of Depression: A Study of Psychiatric Disorder in Women* (Vol. 65). New York: Routledge; Slavich, G. M., Thornton, T., Torres, L. D., Monroe, S. M., & Gotlib, I. H. (2009). Targeted rejection predicts hastened onset of major depression. *Journal of Social and Clinical Psychology, 28*(2), 223.

99 Having a poor social network House, J. S., Landis, K. R., & Umberson, D. (1988). Social relationships and health. *Science, 241*(4865), 540–545; Holt-Lunstad, J., Smith, T. B., & Layton, J. B. (2010). Social relationships and mortality risk: A meta-analytic review. *PLOS Medicine, 7*(7), e1000316.

Chapter 5: Mental Magic Tricks

103 a group of medical residents were each asked to flip a coin 300 times Clark, M. P. A., & Westerberg, B. D. (2009). How random is a coin toss. *Canadian Medical Association Journal, 181,* E306–E308.

103 Statisticians from Stanford University analyzed the physics of coin tossing Diaconi, P., Holmes, S., & Montgomery, R. (2007). Dynamical bias in the coin toss. *SIAM Review, 49,* 211–235.

104 "predicting what your opponent predicts you'll throw" http://www .pleasantmorningbuzz.com/blog/1122061.

105 the first modern texts on psychology Brentano, F. (1995/1874). *Psychology from an Empirical Standpoint.* New York: Routledge; Wundt, W. M. (1904/1874). *Principles of Physiological Psychology* (Vol. 1). London: Sonnenschein.

106 To first demonstrate this penchant for everyday mindreading Heider, F., & Simmel, M. (1944). An experimental study of apparent behavior. *American Journal of Psychology, 57,* 243–259.

107 our tendency to see others in terms of minds guiding behavior Dennett, D. C. (1971). Intentional systems. *Journal of Philosophy, 68,* 87–106.

109 Are we humans alone on the planet Premack, D., & Woodruff, G. (1978). Does the chimpanzee have a theory of mind? *Behavioral and Brain Sciences, 1*(04), 515–526.

110 Very young children watching a Punch and Judy show Dennett, D. C. (1978). Beliefs about beliefs. *Behavioral and Brain Sciences, 1*(04), 568–570.

110 To date, no chimpanzee has shown definitive evidence Call, J., & Tomasello, M. (2008). Does the chimpanzee have a theory of mind? 30 years later. *Trends in Cognitive Sciences, 12*(5), 187–192.

110 converted Dennett's Punch and Judy thought experiment into a real one Wimmer, H., & Perner, J. (1983). Beliefs about beliefs: Representation and constraining function of wrong beliefs in young children's understanding of deception. *Cognition, 13*(1), 103–128; Baron-Cohen, S., Leslie, A. M., & Frith, U. (1985). Does the autistic child have a "theory of mind"? *Cognition, 21*(1), 37–46.

111 The results from many studies provide strong converging evidence Happé, F. G. (1995). The role of age and verbal ability in the theory of mind task performance of subjects with autism. *Child Development, 66*(3), 843–855.

111 younger and younger children also show some evidence of this sort of social skill Buttelmann, D., Carpenter, M., & Tomasello, M. (2009). Eighteen-month-old infants show false belief understanding in an active helping paradigm. *Cognition, 112*(2), 337–342; Kuhlmeier, V., Wynn, K., & Bloom, P. (2003). Attribution of dispositional states by 12-month-olds. *Psychological Science, 14*(5), 402–408.

111 Chimpanzees show evidence of precursors of this ability Cheney, D. L. (2011). Extent and limits of cooperation in animals. *Proceedings of the National Academy of Sciences, 108*, 10902–10909; Call, J., & Tomasello, M. (2008). Does the chimpanzee have a theory of mind? 30 years later. *Trends in Cognitive Sciences, 12*(5), 187–192.

112 our general ability for abstract reasoning and effortful thinking supported by the prefrontal cortex Price, B. H., Daffner, K. R., Stowe, R. M., & Mesilam, M. M. (1990). The comportmental learning disabilities of early frontal lobe damage. *Brain, 113*(5), 1383–1393; Davis, H. L., & Pratt, C. (1995). The development of children's Theory of Mind: The working memory explanation. *Australian Journal of Psychology, 47*, 25–31; Gordon, A. C. L., & Olson, D. R. (1998). The relation between acquisition of a Theory of Mind and the capacity to hold in mind. *Journal of Experimental Child Psychology, 68*, 70–83.

113 identified regions in the lateral prefrontal cortex Goel, V., & Dolan, R. J. (2004). Differential involvement of left prefrontal cortex in inductive and deductive reasoning. *Cognition, 93*(3), B109–B121.

114 Countless fMRI studies of working memory Rottschy, C., et al. (2012). Modelling neural correlates of working memory: A coordinate-based meta-analysis. *NeuroImage, 60*, 830–846.

114 the same lateral frontoparietal regions involved in working memory and reasoning Gray, J. R., Chabris, C. F., & Braver, T. S. (2003). Neural mechanisms of general fluid intelligence. *Nature Neuroscience, 6*(3),

316–322; Lee, K. H., Choi, Y. Y., Gray, J. R., Cho, S. H., Chae, J. H., Lee, S., & Kim, K. (2006). Neural correlates of superior intelligence: Stronger recruitment of posterior parietal cortex. *NeuroImage, 29*(2), 578–586.

114 **why shouldn't it support reasoning about other minds** Indeed an early neuropsychological case study linked lateral prefrontal damage to deficits in Theory of Mind types of tasks; however, it is believed in retrospect that this deficit had more to do with the general difficulty of the task rather than Theory of Mind, per se. Price, B., Daffner, K., Stowe, R., & Mesulam, M. (1990). The comportmental learning disabilities of early frontal lobe damage. *Brain, 113,* 1383–1393; Stone, V. E., Baron-Cohen, S., & Knight, R. T. (1998). Frontal lobe contributions to theory of mind. *Journal of Cognitive Neuroscience, 10*(5), 640–656.

115 **the brain typically handles these two kinds of thinking using very different neural systems** Fletcher, P. C., Happe, F., Frith, U., Baker, S. C., Dolan, R. J., Frackowiak, R. S., & Frith, C. D. (1995). Other minds in the brain: A functional imaging study of "theory of mind" in story comprehension. *Cognition, 57*(2), 109–128.

116 **produced activity in lateral prefrontal regions associated with language and working memory** Rottschy, C., Langner, R., Dogan, I., Reetz, K., Laird, A. R., Schulz, J. B., . . . , & Eickhoff, S. B. (2011). Modelling neural correlates of working memory: A coordinate-based meta-analysis. *NeuroImage. 60,* 830–846; Bavelier, D., Corina, D., Jezzard, P., Padmanabhan, S., Clark, V. P., Karni, A., . . . , & Neville, H. (1997). Sentence reading: A functional MRI study at 4 Tesla. *Journal of Cognitive Neuroscience, 9*(5), 664–686; Turkeltaub, P. E., Gareau, L., Flowers, D. L., Zeffiro, T. A., & Eden, G. F. (2003). Development of neural mechanisms for reading. *Nature Neuroscience, 6*(7), 767–773.

116 **they produced selective activity in the dorsomedial prefrontal cortex (DMPFC) and the temporoparietal junction (TPJ)** Castelli, F., Frith, C., Happé, F., & Frith, U. (2002). Autism, Asperger syndrome and brain mechanisms for the attribution of mental states to animated shapes. *Brain, 125*(8), 1839–1849.

117 **One of my favorite mentalizing studies** St. Jacques, P. L., Conway, M. A., Lowder, M. W., & Cabeza, R. (2011). Watching my mind unfold versus yours: An fMRI study using a novel camera technology to examine neural differences in self-projection of self versus other perspectives. *Journal of Cognitive Neuroscience, 23*(6), 1275–1284.

117 **two things have remained pretty constant** Lieberman, M. D. (2010). Social cognitive neuroscience. In S. T. Fiske, D. T. Gilbert, & G. Lindzey (Eds). *Handbook of Social Psychology,* 5th ed. New York: McGraw-Hill, pp. 143–193; Van Overwalle, F. (2011). A dissociation between social mentalizing and general reasoning. *NeuroImage, 54*(2), 1589–1599.

118 **whenever a person is given a moment of peace in the scanner, between**

cognitive tasks Raichle, M. E., MacLeod, A. M., Snyder, A. Z., Powers, W. J., Gusnard, D. A., & Shulman, G. L. (2001). A default mode of brain function. *Proceedings of the National Academy of Sciences, 98*(2), 676–682.

118 **the same regions that "turn on" when we dream** Braun, A. R., Balkin, T. J., Wesenten, N. J., Carson, R. E., Varga, M., Baldwin, P., . . . , & Herscovitch, P. (1997). Regional cerebral blood flow throughout the sleep-wake cycle. An H2 (15) O PET study. *Brain, 120*(7), 1173–1197; Muzur, A., Pace-Schott, E. F., & Hobson, J. A. (2002). The prefrontal cortex in sleep. *Trends in Cognitive Sciences, 6*(11), 475–481.

118 **our social focus on other people's minds** Spunt, R. P., Meyer, M. L., & Lieberman, M. D. (under review). Social by default: Brain activity at rest facilitates social cognition.

118 **Previous studies have demonstrated that the default network** Harrison, B. J., Pujol, J., López-Solà, M., Hernández-Ribas, R., Deus, J., Ortiz, H., . . . , & Cardoner, N. (2008). Consistency and functional specialization in the default mode brain network. *Proceedings of the National Academy of Sciences, 105*(28), 9781–9786; Spreng, R. N., Mar, R. A., & Kim, A. S. (2009). The common neural basis of autobiographical memory, prospection, navigation, theory of mind, and the default mode: A quantitative meta-analysis. *Journal of Cognitive Neuroscience, 21*(3), 489–510.

118 **it is mostly something that gets in the way, making us more error prone** Anticevic, A., Repovs, G., Shulman, G. L., & Barch, D. M. (2010). When less is more: TPJ and default network deactivation during encoding predicts working memory performance. *NeuroImage, 49*(3), 2638–2648; Li, C. S. R., Yan, P., Bergquist, K. L., & Sinha, R. (2007). Greater activation of the "default" brain regions predicts stop signal errors. *NeuroImage, 38*(3), 640–648.

120 **always for the sake of my doing"** James, W. (1950/1890). *The Principles of Psychology.* New York: Dover.

121 **ran a neuroimaging study on a version of this task, called *Stag Hunt*** Yoshida, W., Seymour, B., Friston, K. J., & Dolan, R. J. (2010). Neural mechanisms of belief inference during cooperative games. *Journal of Neuroscience, 30*(32), 10744–10751.

122 **captured this mentalizing arms race phenomenon** Coricelli, G., & Nagel, R. (2009). Neural correlates of depth of strategic reasoning in medial prefrontal cortex. *Proceedings of the National Academy of Sciences, 106*(23), 9163–9168.

123 **it is linked to the mentalizing system in the brain** Psychologists would probably disagree that guessing 0 represents the most strategic answer in the above scenario. Presumably there would be a mix of nonstrategic, mildly strategic, and very strategic participants, and assessing that mix would give you a higher number than 0 as the optimal answer.

124 **To examine this, we had people lie in a scanner** Falk, E. B., Morelli, S. A., Welbourn, B. L., Dambacher, K., & Lieberman, M. D. (in press). Creating buzz: The neural correlates of effective message propagation. *Psychological Science.*

126 **How effortlessly does our mentalizing system work?** Spunt, R. P., & Lieberman, M. D. (in press). Automaticity, control, and the social brain. In J. Sherman, B. Gawronski, & Y. Trope (Eds.). *Dual Process Theories of the Social Mind.* New York: Guilford; Apperly, I. A., Riggs, K. J., Simpson, A., Chiavarino, C., & Samson, D. (2006). Is belief reasoning automatic? *Psychological Science, 17*(10), 841–844.

126 **like a working memory system, a *social working memory system*** Meyer, M. L., Spunt, R. P., Berkman, E. T., Taylor, S. E., & Lieberman, M. D. (2012). Social working memory: An fMRI study of parametric increases in social cognitive effort. *Proceedings of the National Academy of Sciences, 109,* 1883–1888; Wagner, D. D., Kelley, W. M., & Heatherton, T. F. (2011). Individual differences in the spontaneous recruitment of brain regions supporting mental state understanding when viewing natural social scenes. *Cerebral Cortex, 21*(12), 2788–2796.

126 **to greater increases in mentalizing system activity than easier trials** Mckiernan, K. A., Kaufman, J. N., Kucera-Thompson, J., & Binder, J. R. (2003). A parametric manipulation of factors affecting task-induced deactivation in functional neuroimaging. *Journal of Cognitive Neuroscience, 15*(3), 394–408.

126 **we found that performance on a social working memory task** Dumontheil, I., Jensen, S. G., Wood, N. W., Meyer, M. L., Lieberman, M. D., & Blakemore, S. (under review). Influence of dopamine regulating genes on social working memory.

127 **There is a big difference between having the capacity** Berkman, E., & Lieberman, M. D. (2009). Using neuroscience to broaden emotion regulation: Theoretical and methodological considerations. *Social and Personality Psychology Compass, 3,* 475–493.

127 **We often use our own mind as a proxy** Griffin, D. W., & Ross, L. (1991). Subjective construal, social inference, and human misunderstanding. *Advances in Experimental Social Psychology, 24,* 319–359.

127 **created an elegant paradigm called the *director's task*** Keysar, B., Barr, D. J., Balin, J. A., & Brauner, J. S. (2000). Taking perspective in conversation: The role of mutual knowledge in comprehension. *Psychological Science, 11*(1), 32–38.

128 **What should you do when your partner tells you** Dumontheil, I., Apperly, I. A., & Blakemore, S. J. (2010). Online usage of theory of mind continues to develop in late adolescence. *Developmental Science, 13*(2), 331–338.

Chapter 6: Mirror, Mirror

132 **Some neurons responded to the sight of an object** Rizzolatti, G., Gentilucci, M., Camarda, R. M., Gallese, V., Luppino, G., Matelli, M., & Fogassi, L. (1990). Neurons related to reaching-grasping arm movements in the rostral part of area 6 (area 6aβ). *Experimental Brain Research, 82*(2), 337–350.

132 **changed our fundamental understanding of how we came to be such social creatures** Pellegrino, G. D., Fadiga, L., Fogassi, L., Gallese, V., & Rizzolatti, G. (1992). Understanding motor events: A neurophysiological study. *Experimental Brain Research, 91*(1), 176–180.

133 **Although some psychologists had argued for this kind of perceptual-motor** Prinz, W. (1997). Perception and action planning. *European Journal of Cognitive Psychology, 9*(2), 129–154.

133 **that hitherto remained mysterious and inaccessible to experiments"** Ramachandran, V. S. (2000). Mirror neurons and imitation learning as the driving force behind "the great leap forward" in human evolution. *Edge* website article: http://www.edge.org/3rd_culture/ramachandran /ramachandran_p1.html.

133 **our capacity for language, culture, imitation, mindreading, and empathy** Arbib, M. A. (2005). From monkey-like action recognition to human language: An evolutionary framework for neurolinguistics. *Behavioral and Brain Sciences, 28*(02), 105–124; Molenberghs, P., Cunnington, R., & Mattingley, J. B. (2009). Is the mirror neuron system involved in imitation? A short review and meta-analysis. *Neuroscience & Biobehavioral Reviews, 33*(7), 975–980; Blakeslee, S. (2006). Cells that read minds. *New York Times,* January 10, p. 1; Fabrega Jr., H. (2005). Biological evolution of cognition and culture: Off Arbib's mirror-neuron system stage? *Behavioral and Brain Sciences, 28*(02), 131–132; Gallese, V. (2001). The shared manifold hypothesis. From mirror neurons to empathy. *Journal of Consciousness Studies, 8*(5-7), 33–50.

134 **allowing us to keep more abstract ideas in mind at the same time** Coolidge, F. L., & Wynn, T. (2005). Working memory, its executive functions, and the emergence of modern thinking. *Cambridge Archaeological Journal, 15*(1), 5–26.

134 **force behind the 'great leap forward' in human evolution"** Ramachandran, V. S. (2000). Mirror neurons and imitation learning as the driving force behind "the great leap forward" in human evolution. *Edge* website article: http://www.edge.org/3rd_culture/ramachandran /ramachandran_p1. html.

135 **first evidence regarding the presence of a mirror neuron system in humans** Iacoboni, M., Woods, R. P., Brass, M., Bekkering, H., Mazziotta,

J. C., & Rizzolatti, G. (1999). Cortical mechanisms of human imitation. *Science, 286*(5449), 2526–2528.

136 **temporarily "frazzles" the neurons in that area such that the region is essentially taken offline** Heiser, M., Iacoboni, M., Maeda, F., Marcus, J., & Mazziotta, J. C. (2003). The essential role of Broca's area in imitation. *European Journal of Neuroscience, 17*(5), 1123–1128.

136 **imitating the fingering required to make a set of guitar chords that they were shown** Buccino, G., Vogt, S., Ritzl, A., Fink, G. R., Zilles, K., Freund, H. J., & Rizzolatti, G. (2004). Neural circuits underlying imitation learning of hand actions: An event-related fMRI study. *Neuron, 42*(2), 323–334.

138 **Gordon's account of the second route** Ross, L., Greene, D., & House, P. (1977). The "false consensus effect": An egocentric bias in social perception and attribution processes. *Journal of Experimental Social Psychology, 13*(3), 279–301; Ames, D. R. (2004). Inside the mind reader's tool kit: Projection and stereotyping in mental state inference. *Journal of Personality and Social Psychology, 87*(3), 340.

138 **mirror neurons are the neural implementation of Simulation theory** Gallese, V., & Goldman, A. (1998). Mirror neurons and the simulation theory of mind-reading. *Trends in Cognitive Sciences, 2*(12), 493–501.

138 **"the fundamental mechanism that allows us a direct experiential grasp** Gallese, V., Keysers, C., & Rizzolatti, G. (2004). A unifying view of the basis of social cognition. *Trends in Cognitive Sciences, 8*(9), 396–403; Gordon actually foreshadowed this idea two decades earlier writing that an "interesting possibility for . . . practical simulation is a prepackaged 'module' called upon automatically in the perception of other human beings": Gordon, R. M. (2007). Folk psychology as simulation. *Mind & Language, 1*(2), 158–171.

139 **My brain is mirroring your brain** Rizzolatti, G., & Sinigaglia, C. (2010). The functional role of the parieto-frontal mirror circuit: Interpretations and misinterpretations. *Nature Reviews Neuroscience, 11*(4), 264–274.

140 **found mirror neurons that fit the bill** Kohler, E., Keysers, C., Umilta, M. A., Fogassi, L., Gallese, V., & Rizzolatti, G. (2002). Hearing sounds, understanding actions: Action representation in mirror neurons. *Science, 297*(5582), 846–848.

140 **highlighted an important limitation of the sight-sound study** Hickok, G. (2009). Eight problems for the mirror neuron theory of action understanding in monkeys and humans. *Journal of Cognitive Neuroscience, 21*(7), 1229–1243.

141 **that the mirror neurons respond to actions that involve objects that cannot be seen** Umilta, M. A., Kohler, E., Gallese, V., Fogassi, L., Fadiga, L., Keysers, C., & Rizzolatti, G. (2001). I know what you are doing: A neurophysiological study. *Neuron, 31*(1), 155–166.

141 **Humans are certainly capable of seeing an object** Lee, H., Simpson, G. V., Logothetis, N. K., & Rainer, G. (2005). Phase locking of single neuron activity to theta oscillations during working memory in monkey extrastriate visual cortex. *Neuron, 45*(1), 147–156.

141 **She suggests that the purpose of mirror neurons** Heyes, C. (2010). Mesmerising mirror neurons. *NeuroImage, 51*(2), 789–791.

142 **Heyes designed a clever *countermirroring* procedure** Catmur, C., Walsh, V., & Heyes, C. (2007). Sensorimotor learning configures the human mirror system. *Current Biology, 17*(17), 1527–1531; Catmur, C., Mars, R. B., Rushworth, M. F., & Heyes, C. (2011). Making mirrors: Premotor cortex stimulation enhances mirror and counter-mirror motor facilitation. *Journal of Cognitive Neuroscience, 23*(9), 2352–2362.

142 **Another study examined how the mirror system responds** Newman-Norlund, R. D., van Schie, H. T., van Zuijlen, A. M., & Bekkering, H. (2007). The mirror neuron system is more active during complementary compared with imitative action. *Nature Neuroscience, 10*(7), 817–818.

144 **If we look at brains at rest** Fox, M. D., Snyder, A. Z., Vincent, J. L., Corbetta, M., Van Essen, D. C., & Raichle, M. E. (2005). The human brain is intrinsically organized into dynamic, anticorrelated functional networks. *Proceedings of the National Academy of Sciences of the United States of America, 102*(27), 9673–9678.

144 **If he answers, "In order to have a drink"** Spunt, R. P., & Lieberman, M. D. (in press). Automaticity, control, and the social brain. In J. Sherman, B. Gawronski, & Y. Trope (Eds.). *Dual Process Theories of the Social Mind.* New York: Guilford.

145 **systematically investigated these distinctions** Vallacher, R. R., & Wegner, D. M. (1987). What do people think they're doing? Action identification and human behavior. *Psychological Review, 94*(1), 3.

145 **are more likely to focus on the thoughts they are trying to convey** Carver, C. S. (1979). A cybernetic model of self-attention processes. *Journal of Personality and Social Psychology, 37*(8), 1251.

146 **explain another person's high-level reasons for wanting the light on** Jacob, P., & Jeannerod, M. (2005). The motor theory of social cognition: A critique. *Trends in Cognitive Sciences, 9*(1).

147 **figuring out the how, what, and why of other people's behavior** Spunt, R. P., & Lieberman, M. D. (2012). Dissociating modality-specific and supramodal neural systems for action understanding. *Journal of Neuroscience, 32,* 3575–3583; Spunt, R. P., & Lieberman, M. D. (2012). An integrative model of the neural systems supporting the comprehension of observed emotional behavior. *NeuroImage, 59,* 3050–3059; Spunt, R. P., Falk, E. B., & Lieberman, M. D. (2010). Dissociable neural systems support retrieval of "how" and "why" action knowledge. *Psychological Science, 21,* 1593–1598; Spunt, R .P., Satpute, A. B., & Lieberman, M. D.

(2011). Identifying the what, why, and how of an observed action: An fMRI study of mentalizing and mechanizing during action observation. *Journal of Cognitive Neuroscience, 23,* 63–74; Brass, M., Schmitt, R. M., Spengler, S., & Gergely, G. (2007). Investigating action understanding: Inferential processes versus action simulation. *Current Biology, 17*(24), 2117–2121; de Lange, F. P., Spronk, M., Willems, R. M., Toni, I., & Bekkering, H. (2008). Complementary systems for understanding action intentions. *Current Biology, 18*(6), 454–457; Noordzij, M. L., Newman-Norlund, S. E., De Ruiter, J. P., Hagoort, P., Levinson, S. C., & Toni, I. (2009). Brain mechanisms underlying human communication. *Frontiers in Human Neuroscience, 3,* 14.

148 **there was increased mirror system activity** Spunt, R. P., & Lieberman, M. D. (2012). Dissociating modality-specific and supramodal neural systems for action understanding. *Journal of Neuroscience, 32,* 3575–3583.

148 **this was still the case when participants were distracted** Spunt, R. P., & Lieberman, M. D. (2013). The busy social brain: Evidence for automaticity and control in the neural systems supporting social cognition and action understanding. *Psychological Science, 24,* 80–86.

149 **"blooming, buzzing confusion"** James, W. (1890/1950). *The Principles of Psychology.* New York: Dover, p. 462.

Chapter 7: Peaks and Valleys

152 **The word *empathy* was introduced into the English language** Titchener, E. B. (1909). *Lectures on the Experimental Psychology of Thought-Processes.* New York: Macmillan.

152 **something like a first-person experience from the object's perspective** The modern use of the term to refer to appreciating another person's experience can be traced to Husserl's *Ideen* and his student Edith Stein's dissertation published as Stein, E. (1989/1916). *On the Problem of Empathy.* Washington D.C.: ICS Publications.

152 **There are at least three kinds of psychological processes** Zaki, J., & Ochsner, K. (2012). The neuroscience of empathy: Progress, pitfalls and promise. *Nature Neuroscience, 15,* 675-680.

153 **individuals watched others receive shocks to their hands or feet** Avenanti, A., Bueti, D., Galati, G., & Aglioti, S. M. (2005). Transcranial magnetic stimulation highlights the sensorimotor side of empathy for pain. *Nature Neuroscience, 8*(7), 955–960.

153 **muscles in our own faces immediately mimic the expression in subtle ways** Dimberg, U., Thunberg, M., & Elmehed, K. (2000). Unconscious facial reactions to emotional facial expressions. *Psychological Science, 11*(1), 86–89.

153 **if a person is unable to mimic those facial expressions** Neal, D. T., & Chartrand, T. L. (2011). Embodied emotion perception amplifying and dampening facial feedback modulates emotion perception accuracy. *Social Psychological and Personality Science, 2*(6), 673–678.

153 **Given that the mirror system is involved in understanding** Wicker, B., Keysers, C., Plailly, J., Royet, J. P., Gallese, V., & Rizzolatti, G. (2003). Both of us disgusted in *my* insula: The common neural basis of seeing and feeling disgust. *Neuron, 40*(3), 655–664; Carr, L., Iacoboni, M., Dubeau, M. C., Mazziotta, J. C., & Lenzi, G. L. (2003). Neural mechanisms of empathy in humans: A relay from neural systems for imitation to limbic areas. *Proceedings of the National Academy of Sciences, 100*(9), 5497–5502.

154 **trying to understand why someone is experiencing a particular emotion** Spunt, R. P., & Lieberman, M. D. (2012). An integrative model of the neural systems supporting the comprehension of observed emotional behavior. *NeuroImage, 59,* 3050–3059; Mar, R. A. (2011). The neural bases of social cognition and story comprehension. *Annual Review of Psychology, 62,* 103–134; Singer, T., Seymour, B., O'Doherty, J., Kaube, H., Dolan, R. J., & Frith, C. D. (2004). Empathy for pain involves the affective but not sensory components of pain. *Science, 303*(5661), 1157–1162.

156 **affect matching can sometimes lead to avoidance behavior** Batson, C. D. (1991). *The Altruism Question: Toward a Social-Psychological Answer.* Hillsdale, NJ: Lawrence Erlbaum Associates.

156 **empathy occurs only when there is an appropriate emotional response** Ibid.

156 **Given that nearly all of the follow-ups to Singer's seminal work have focused on empathy** Fan, Y., Duncan, N. W., de Greck, M., & Northoff, G. (2011). Is there a core neural network in empathy? An fMRI based quantitative meta-analysis. *Neuroscience & Biobehavioral Reviews, 35*(3), 903–911.

156 **almost none of the studies that have been done have linked neural responses** Hein, G., Silani, G., Preuschoff, K., Batson, C. D., & Singer, T. (2010). Neural responses to ingroup and outgroup members' suffering predict individual differences in costly helping. *Neuron, 68*(1), 149–160.

156 **Sylvia Morelli, Lian Rameson, and I ran an fMRI study** Morelli, S. A., Rameson, L. T., & Lieberman, M. D. (in press). The neural components of empathy: Predicting daily prosocial behavior. *Social Cognitive and Affective Neuroscience.*

158 **the hot area of study in the next ten years** Cf. Moll, J., Zahn, R., de Oliveira-Souza, R., Bramati, I. E., Krueger, F., Tura, B., . . . , & Grafman, J. (2011). Impairment of prosocial sentiments is associated with frontopolar and septal damage in frontotemporal dementia. *NeuroImage, 54*(2), 1735–1742; Krueger, F., McCabe, K., Moll, J., Kriegeskorte, N., Zahn, R., Strenziok, M., . . . , & Grafman, J. (2007). Neural correlates

of trust. *Proceedings of the National Academy of Sciences, 104*(50), 20084–20089; Inagaki, T. K., & Eisenberger, N. I. (2012). Neural correlates of giving support to a loved one. *Psychosomatic Medicine, 74,* 3–7.

158 **the dorsomedial prefrontal cortex (DMPFC)—the CEO of the brain's mentalizing system** Andy, O. J., & Stephan, H. (1966). Septal nuclei in primate phylogeny: A quantitative investigation. *Journal of Comparative Neurology, 126*(2), 157–170; Sesack, S. R., Deutch, A. Y., Roth, R. H., & Bunney, B. S. (1989). Topographical organization of the efferent projections of the medial prefrontal cortex in the rat: An anterograde tract-tracing study with *Phaseolus vulgaris leucoagglutinin. Journal of Comparative Neurology, 290*(2), 213–242.

159 **When a rat pressed the lever** Olds, J., & Milner, P. (1954). Positive reinforcement produced by electrical stimulation of septal area and other regions of rat brain. *Journal of Comparative and Physiological Psychology, 47*(6), 419.

159 **Two decades later a similar study was conducted** Heath, R. G. (1972). Pleasure and brain activity in man. *Journal of Nervous and Mental Disease, 154*(363), 9.

159 **At the same time that researchers were linking the septal area to reward** Brady, J. V., & Nauta, W. J. (1953). Subcortical mechanisms in emotional behavior: Affective changes following septal forebrain lesions in the albino rat. *Journal of Comparative and Physiological Psychology, 46*(5), 339.

159 **Lesion studies in rats, mice, and rabbits suggest** Carlson, N. R., & Thomas, G. J. (1968). Maternal behavior of mice with limbic lesions. *Journal of Comparative and Physiological Psychology, 66*(3p1), 731; Cruz, M. L., & Beyer, C. (1972). Effects of septal lesions on maternal behavior and lactation in the rabbit. *Physiology & Behavior, 9*(3), 361–365; Slotnick, B. M., & Nigrosh, B. J. (1975). Maternal behavior of mice with cingulate cortical, amygdala, or septal lesions. *Journal of Comparative and Physiological Psychology, 88*(1), 118.

160 **one way to reconcile the findings is to characterize the septal area as shifting the balance** Inagaki, T. K., & Eisenberger, N. I. (2012). Neural correlates of giving support to a loved one. *Psychosomatic Medicine, 74,* 3–7.

160 **The septal area is rich in oxytocin receptors** Insel, T. R., Gelhard, R., & Shapiro, L. E. (1991). The comparative distribution of forebrain receptors for neurohypophyseal peptides in monogamous and polygamous mice. *Neuroscience, 43*(2), 623–630.

160 **Among rodents, pups who receive more parental care** Lukas, M., Bredewold, R., Neumann, I. D., & Veenema, A. H. (2010). Maternal separation interferes with developmental changes in brain vasopressin and oxytocin receptor binding in male rats. *Neuropharmacology, 58*(1),

78–87; Francis, D. D., Champagne, F. C., & Meaney, M. J. (2001). Variations in maternal behaviour are associated with differences in oxytocin receptor levels in the rat. *Journal of Neuroendocrinology, 12*(12), 1145–1148.

162 **Just two years after the first Sally-Anne test for Theory of Mind** Baron-Cohen, S., Leslie, A. M., & Frith, U. (1985). Does the autistic child have a "theory of mind"? *Cognition, 21*(1), 37–46.

163 **Subsequent studies have demonstrated other mentalizing deficits** Baron-Cohen, S., O'Riordan, M., Stone, V., Jones, R., & Plaisted, K. (1999). Recognition of faux pas by normally developing children and children with Asperger syndrome or high-functioning autism. *Journal of Autism and Developmental Disorders, 29,* 407–418; White, S. J., Hill, E. L., Happé, F., & Frith, U. (2009). Revisiting the Strange Stories: Revealing mentalizing impairments in autism. *Child Development, 80,* 1097–1117.

163 **Autistic individuals are also much less likely to describe** Heider, F., & Simmel, M. (1944). An experimental study of apparent behavior. *American Journal of Psychology, 57,* 243–259; Klin, A. (2003). Attributing social meaning to ambiguous visual stimuli in higher-functioning autism and Asperger syndrome: The social attribution task. *Journal of Child Psychology and Psychiatry, 41*(7), 831–846.

165 **The Sally-Anne test isn't the only or the hardest test of Theory of Mind** Frith, U., & Happé, F. (1994). Autism: Beyond "theory of mind." *Cognition, 50*(1), 115–132.

165 **Uta Frith gave a group of autistic children** Shah, A., & Frith, U. (1983). An islet of ability in autistic children: A research note. *Journal of Child Psychology and Psychiatry, 24*(4), 613–620.

166 **Autism is associated with a deficit in focusing on high-level meaning** Frith, U., & Happé, F. (1994). Autism: Beyond "theory of mind." *Cognition, 50*(1), 115–132.

166 **the two deficits do not always go hand in hand in autism** Ibid.; Spunt, R. P., Meyer, M. L., & Lieberman, M. D. (under review). Social by default: Brain activity at rest facilitates social cognition; cf. Baron-Cohen, S. (2009). Autism: The Empathizing-Systemizing (E-S) Theory. *Annals of the New York Academy of Sciences, 1156*(1), 68–80.

167 **Multiple studies have shown that training can lead to sizable gains** Hadwin, J., Baron-Cohen, S., Howlin, P., & Hill, K. (1997). Does teaching theory of mind have an effect on the ability to develop conversation in children with autism? *Journal of Autism and Developmental Disorders, 27*(5), 519–537; Ozonoff, S., & Miller, J. N. (1995). Teaching theory of mind: A new approach to social skills training for individuals with autism. *Journal of Autism and Developmental Disorders, 25*(4), 415–433.

168 **We know this because children who are born deaf** Peterson, C. C., & Siegal, M. (1999). Deafness, conversation and theory of mind. *Journal*

of Child Psychology and Psychiatry, 36(3), 459–474; Peterson, C. C., & Siegal, M. (1999). Representing inner worlds: Theory of mind in autistic, deaf, and normal hearing children. *Psychological Science, 10*(2), 126–129; Peterson, C. C., & Siegal, M. (2002). Insights into theory of mind from deafness and autism. *Mind & Language, 15*(1), 123–145.

168 **During the second year, these children tend to ignore others** Adrien, J. L., Lenoir, P., Martineau, J., Perrot, A., Hameury, L., Larmande, C., & Sauvage, D. (1993). Blind ratings of early symptoms of autism based upon family home movies. *Journal of the American Academy of Child & Adolescent Psychiatry, 32*(3), 617–626; Klin, A., Volkmar, F. R., & Sparrow, S. S. (1992). Autistic social dysfunction: Some limitations of the theory of mind hypothesis. *Journal of Child Psychology and Psychiatry, 33*(5), 861–876.

169 **experimenters have asked children to imitate various behaviors and hand gestures** DeMyer, M. K., Alpern, G. D., Barton, S., DeMyer, W. E., Churchill, D. W., Hingtgen, J. N., . . . , & Kimberlin, C. (1972). Imitation in autistic, early schizophrenic, and non-psychotic subnormal children. *Journal of Autism and Developmental Disorders, 2*(3), 264–287.

169 **Children diagnosed with autism consistently perform worse** Williams, J. H., Whiten, A., & Singh, T. (2004). A systematic review of action imitation in autistic spectrum disorder. *Journal of Autism and Developmental Disorders, 34*(3), 285–299.

169 **Once the mirror system was clearly linked with imitation** Nishitani, N., Avikainen, S., & Hari, R. (2004). Abnormal imitation-related cortical activation sequences in Asperger's syndrome. *Annals of Neurology, 55*(4), 558–562; Oberman, L. M., Hubbard, E. M., McCleery, J. P., Altshuler, E. L., Ramachandran, V. S., & Pineda, J. A. (2005). EEG evidence for mirror neuron dysfunction in autism spectrum disorders. *Cognitive Brain Research, 24,* 190–198; Dapretto, M., Davies, M. S., Pfeifer, J. H., Scott, A. A., Sigman, M., Bookheimer, S. Y., & Iacoboni, M. (2005). Understanding emotions in others: Mirror neuron dysfunction in children with autism spectrum disorders. *Nature Neuroscience, 9*(1), 28–30; Williams, J. H., Waiter, G. D., Gilchrist, A., Perrett, D. I., Murray, A. D., & Whiten, A. (2006). Neural mechanisms of imitation and mirror neuron functioning in autistic spectrum disorder. *Neuropsychologia, 44*(4), 610–621; Ramachandran, V. S., & Oberman, L. M. (2006). Broken mirrors: A theory of autism. *Scientific American, 16,* 62–69; Gallese, V. (2006). Intentional attunement: A neurophysiological perspective on social cognition and its disruption in autism. *Brain Research, 1079*(1), 15–24.

169 **individuals with autism produced mu suppression only when they performed hand actions themselves** Oberman, L. M., Hubbard, E. M., McCleery, J. P., Altshuler, E. L., Ramachandran, V. S., & Pineda, J. A.

(2005). EEG evidence for mirror neuron dysfunction in autism spectrum disorders. *Cognitive Brain Research, 24,* 190–198.

169 **the conclusion that mirror system activity differs in the autistic sample is unwarranted** Nieuwenhuis, S., Forstmann, B. U., & Wagenmakers, E. J. (2011). Erroneous analyses of interactions in neuroscience: A problem of significance. *Nature Neuroscience, 14*(9), 1105–1107.

169 **One fMRI study found that individuals with autism** Dapretto, M., Davies, M. S., Pfeifer, J. H., Scott, A. A., Sigman, M., Bookheimer, S. Y., & Iacoboni, M. (2005). Understanding emotions in others: Mirror neuron dysfunction in children with autism spectrum disorders. *Nature Neuroscience, 9*(1), 28–30.

170 **The other study found that this group produced decreased activity** Williams, J. H., Waiter, G. D., Gilchrist, A., Perrett, D. I., Murray, A. D., & Whiten, A. (2006). Neural mechanisms of imitation and mirror neuron functioning in autistic spectrum disorder. *Neuropsychologia, 44*(4), 610–621.

170 **Multiple studies have shown roughly equal levels of mu suppression** Fan, Y. T., Decety, J., Yang, C. Y., Liu, J. L., & Cheng, Y. (2010). Unbroken mirror neurons in autism spectrum disorders. *Journal of Child Psychology and Psychiatry, 51*(9), 981–988; Raymaekers, R., Wiersema, J. R., & Roeyers, H. (2009). EEG study of the mirror neuron system in children with high functioning autism. *Brain Research, 1304,* 113–121.

170 **a number of fMRI studies have shown equivalent or enhanced mirror system** Dinstein, I., Thomas, C., Humphreys, K., Minshew, N., Behrmann, M., & Heeger, D. J. (2010). Normal movement selectivity in autism. *Neuron, 66*(3), 461–469; Marsh, L. E., & Hamilton, A. F. D. C. (2011). Dissociation of mirroring and mentalising systems in autism. *NeuroImage, 56*(3), 1511–1519; Martineau, J., Andersson, F., Barthélémy, C., Cottier, J. P., & Destrieux, C. (2010). Atypical activation of the mirror neuron system during perception of hand motion in autism. *Brain Research, 1320,* 168–175.

170 **imitation performance cannot be so easily equated** Southgate, V., & Hamilton, A. F. D. C. (2008). Unbroken mirrors: Challenging a theory of autism. *Trends in Cognitive Sciences, 12*(6), 225–229.

171 **nonautistic individuals take longer to perform the incompatible movement** Bird, G., Leighton, J., Press, C., & Heyes, C. (2007). Intact automatic imitation of human and robot actions in autism spectrum disorders. *Proceedings of the Royal Society B: Biological Sciences, 274*(1628), 3027–3031.

171 **Another group using a different automatic imitation paradigm** Spengler, S., Bird, G., & Brass, M. (2010). Hyperimitation of actions is related to reduced understanding of others' minds in autism spectrum conditions. *Biological Psychiatry, 68*(12), 1148–1155.

173 **We** tend to assume that outsides and insides match Gilbert, D. T., & Malone, P. S. (1995). The correspondence bias. *Psychological Bulletin, 117*(1), 21.

174 **the** intense **world hypothesis of autism** Markram, H., Rinaldi, T., & Markram, K. (2007). The intense world syndrome: An alternative hypothesis for autism. *Frontiers in Neuroscience, 1*(1), 77–96; Markram, K., & Markram, H. (2010). The intense world theory: A unifying theory of the neurobiology of autism. *Frontiers in Human Neuroscience, 4,* 1–29.

174 **People are loud confusing creatures. . . . And they expect me to add eye** contact? http://nolongerinabox.wordpress.com/2012/09/19/on-eye-contact/.

175 **In humans, the amygdala seems particularly responsive** Adolphs, R., Baron-Cohen, S., & Tranel, D. (2002). Impaired recognition of social emotions following amygdala damage. *Journal of Cognitive Neuroscience, 14*(8), 1264–1274.

175 **And while the amygdala does respond to intense positive and negative cues** Small, D. M., Gregory, M. D., Mak, Y. E., Gitelman, D., Mesulam, M. M., & Parrish, T. (2003). Dissociation of neural representation of intensity and affective valuation in human gustation. *Neuron, 39*(4), 701.

175 **Even subliminally presented fearful faces** Morris, J. S., Öhman, A., & Dolan, R. J. (1999). A subcortical pathway to the right amygdala mediating "unseen" fear. *Proceedings of the National Academy of Sciences, 96*(4), 1680–1685; Whalen, P. J., Rauch, S. L., Etcoff, N. L., McInerney, S. C., Lee, M. B., & Jenike, M. A. (1998). Masked presentations of emotional facial expressions modulate amygdala activity without explicit knowledge. *Journal of Neuroscience, 18*(1), 411–418.

175 **The best initial evidence for an amygdala-autism link** Baron-Cohen, S., Ring, H. A., Bullmore, E. T., Wheelwright, S., Ashwin, C., & Williams, S. C. (2000). The amygdala theory of autism. *Neuroscience & Biobehavioral Reviews, 24*(3), 355–364; Critchley, H. D., Daly, E. M., Bullmore, E. T., Williams, S. C., Van Amelsvoort, T., Robertson, D. M., . . . , & Murphy, D. G. (2000). The functional neuroanatomy of social behaviour changes in cerebral blood flow when people with autistic disorder process facial expressions. *Brain, 123*(11), 2203–2212; Pierce, K., Müller, R. A., Ambrose, J., Allen, G., & Courchesne, E. (2001). Face processing occurs outside the fusiform face area in autism: Evidence from functional MRI. *Brain, 124*(10), 2059–2073.

175 **When these findings were combined with the fact that amygdala damage in nonhuman primates** Bachevalier, J. (1991). An animal model for childhood autism: Memory loss and socioemotional disturbances following neonatal damage to the limbic system in monkeys. *Advances in Neuropsychiatry and Psychopharmacology, 1,* 129–140.

175 **Children with autism actually have larger amygdalae** Amaral, D. G., Schumann, C. M., & Nordahl, C. W. (2008). Neuroanatomy of autism. *Trends in Neurosciences, 31*(3), 137–145.

175 **This has been seen in children as young as two to four years old** Mosconi, M. W., Cody-Hazlett, H., Poe, M. D., Gerig, G., Gimpel-Smith, R., & Piven, J. (2009). Longitudinal study of amygdala volume and joint attention in 2-to 4-year-old children with autism. *Archives of General Psychiatry, 66*(5), 509; Schumann, C. M., Hamstra, J., Goodlin-Jones, B. L., Lotspeich, L. J., Kwon, H., Buonocore, M. H., . . . , & Amaral, D. G. (2004). The amygdala is enlarged in children but not adolescents with autism; the hippocampus is enlarged at all ages. *Journal of Neuroscience, 24*(28), 6392–6401.

175 **Seeing that Einstein had an abnormally large parietal lobe** Witelson, S. F., Kigar, D. L., & Harvey, T. (1999). The exceptional brain of Albert Einstein. *Lancet* (London, England), *353*(9170), 2149–2153.

176 **a sign that they may be overwhelmed by the environment** Juranek, J., Filipek, P. A., Berenji, G. R., Modahl, C., Osann, K., & Spence, M. A. (2006). Association between amygdala volume and anxiety level: Magnetic resonance imaging (MRI) study in autistic children. *Journal of Child Neurology, 21*(12), 1051–1058.

176 **Autistic children also show enhanced threat detection** Krysko, K. M., & Rutherford, M. D. (2009). A threat-detection advantage in those with autism spectrum disorders. *Brain and Cognition, 69*(3), 472–480; Kleinhans, N., Johnson, L., Richards, T., Mahurin, R., Greenson, J., Dawson, G., & Aylward, E. (2009). Reduced neural habituation in the amygdala and social impairments in autism spectrum disorders. *American Journal of Psychiatry, 166*(4), 467–475.

176 **increased amygdala volume at age three** Munson, J., Dawson, G., Abbott, R., Faja, S., Webb, S. J., Friedman, S. D., . . . , & Dager, S. R. (2006). Amygdalar volume and behavioral development in autism. *Archives of General Psychiatry, 63*(6), 686.

176 **the visual pathways that feed potential threat information to the amygdala** Samson, F., Mottron, L., Soulières, I., & Zeffiro, T. A. (2011). Enhanced visual functioning in autism: An ALE meta-analysis. *Human Brain Mapping, 33*, 1553-1581.

176 **Some evidence also suggests that autistic individuals** Baron-Cohen, S., Ashwin, E., Ashwin, C., Tavassoli, T., & Chakrabarti, B. (2009). Talent in autism: Hyper-systemizing, hyper-attention to detail and sensory hypersensitivity. *Philosophical Transactions of the Royal Society B: Biological Sciences, 364*(1522), 1377–1383; Blakemore, S. J., Tavassoli, T., Calò, S., Thomas, R. M., Catmur, C., Frith, U., & Haggard, P. (2006). Tactile sensitivity in Asperger syndrome. *Brain and Cognition, 61*(1), 5–13; Crane, L., Goddard, L., & Pring, L. (2009). Sensory processing in adults

with autism spectrum disorders. *Autism, 13*(3), 215–228; Khalfa, S., Bruneau, N., Rogé, B., Georgieff, N., Veuillet, E., Adrien, J. L., . . . , & Collet, L. (2004). Increased perception of loudness in autism. *Hearing Research, 198*(1), 87–92; Kern, J. K., Trivedi, M. H., Garver, C. R., Grannemann, B. D., Andrews, A. A., Savla, J. S., . . . , & Schroeder, J. L. (2006). The pattern of sensory processing abnormalities in autism. *Autism, 10*(5), 480–494.

176 **When you or I see a face** Neumann, D., Spezio, M. L., Piven, J., & Adolphs, R. (2006). Looking you in the mouth: Abnormal gaze in autism resulting from impaired top-down modulation of visual attention. *Social Cognitive and Affective Neuroscience, 1*(3), 194–202.

176 **Nonautistics spend nearly twice as much time looking at the eyes** Pelphrey, K. A., Sasson, N. J., Reznick, J. S., Paul, G., Goldman, B. D., & Piven, J. (2002). Visual scanning of faces in autism. *Journal of Autism and Developmental Disorders, 32*(4), 249–261; Neumann, D., Spezio, M. L., Piven, J., & Adolphs, R. (2006). Looking you in the mouth: Abnormal gaze in autism resulting from impaired top-down modulation of visual attention. *Social Cognitive and Affective Neuroscience, 1*(3), 194–202.

176 **These differences in social gazing (that is, how we look at faces)** Dalton, K. M., Nacewicz, B. M., Johnstone, T., Schaefer, H. S., Gernsbacher, M. A., Goldsmith, H. H., . . . , & Davidson, R. J. (2005). Gaze fixation and the neural circuitry of face processing in autism. *Nature Neuroscience, 8*(4), 519–526.

Chapter 8: Trojan Horse Selves

181 **A few decades later, J. J. Becher published *Physica Subterranea*** Becher, J. J. (1669). *Physica subterranea.* Frankfurt.

182 **In 1970, Gordon Gallup made a mirror available** Gallup, G. G. (1970). Chimpanzees: Self-recognition. *Science, 167*(3914), 86–87.

183 **Paralleling work linking social interaction and self-awareness in humans** Gallup, G. G. (1977). Self-recognition in primates: A comparative approach to the bidirectional properties of consciousness. *American Psychologist, 32*(5), 329.

183 **Not without some controversy, results like these** Plotnik, J. M., de Waal, F. B., & Reiss, D. (2006). Self-recognition in an Asian elephant. *Proceedings of the National Academy of Sciences, 103*(45), 17053–17057; Reiss, D., & Marino, L. (2001). Mirror self-recognition in the bottlenose dolphin: A case of cognitive convergence. *Proceedings of the National Academy of Sciences, 98*(10), 5937–5942.

183 **In addition, the parietal region that responds** Lieberman, M. D. (2007). Social cognitive neuroscience: A review of core processes. *Annual Review of Psychology, 58*, 259–289.

184 **Long before Descartes, the Oracle at Delphi urged all to "know thyself"**
Baumeister, R. F. (1986). *Identity: Cultural Change and the Struggle for Self.* New York: Oxford University Press, p. 153.

184 **Bill Kelley, Todd Heatherton, and Neil Macrae, prominent social neuroscientists** Kelley, W. M., Macrae, C. N., Wyland, C. L., Caglar, S., Inati, S., & Heatherton, T. F. (2002). Finding the self? An event-related fMRI study. *Journal of Cognitive Neuroscience, 14*(5), 785–794.

186 **The medial prefrontal region that was observed** Denny, B. T., Kober, H., Wager, T. D., & Ochsner, K. N. (2012). A meta-analysis of functional neuroimaging studies of self and other judgments reveals a spatial gradient for mentalizing in medial prefrontal cortex. *Journal of Cognitive Neuroscience, 24*(8), 1742–1752.

186 **the MPFC was observed in 94 percent** Lieberman, M. D. (2010). Social cognitive neuroscience. In S. T. Fiske, D. T. Gilbert, & G. Lindzey (Eds). *Handbook of Social Psychology*, 5th ed. New York: McGraw-Hill, pp. 143–193.

187 **only our closer primate relatives** Tsujimoto, S., Genovesio, A., & Wise, S. P. (2011). Frontal pole cortex: Encoding ends at the end of the endbrain. *Trends in Cognitive Sciences, 15*(4), 169–176; Preuss, T. M., & Goldman-Rakic, P. S. (1991). Myelo- and cytoarchitecture of the granular frontal cortex and surrounding regions in the strepsirhine primate Galago and the anthropoid primate Macaca. *Journal of Comparative Neurology, 310*(4), 429–474.

188 **Neuroanatomist Katarina Semendeferi examined the size of BA10** Semendeferi, K., Armstrong, E., Schleicher, A., Zilles, K., & Van Hoesen, G. W. (2001). Prefrontal cortex in humans and apes: A comparative study of area 10. *American Journal of Physical Anthropology, 114*(3), 224–241.

188 **BA10 is less densely populated with neurons** Semendeferi, K., Teffer, K., Buxhoeveden, D. P., Park, M. S., Bludau, S., Amunts, K., . . . , & Buckwalter, J. (2011). Spatial organization of neurons in the frontal pole sets humans apart from great apes. *Cerebral Cortex, 21*(7), 1485–1497.

188 **each "represents the unique, the very special** Hesse, H. (1923). *Demian.* New York: Boni & Liverright.

191 **The generally accepted rule is pink for the boys, and blue for the girls** From *Earnshaw's Infants' Department* in June 1918, as quoted from Smithsonian.com: Jeanne Maglaty, "When Did Girls Start Wearing Pink?" April 8, 2011.

192 **learning about ourselves in the real world** Mead, G. H. (1934). *Mind, Self, and Society from the Standpoint of a Social Behaviorist* (C. W. Morris, Ed.). Chicago: University of Chicago; Cooley, C. H. (1902). *Human Nature and the Social Order.* New York: Scribner.

193 **We asked young adolescents (that is, thirteen-year-olds)** Pfeifer, J. H., Masten, C. L., Borofsky, L. A., Dapretto, M., Fuligni, A. J., & Lieberman,

M. D. (2009). Neural correlates of direct and reflected self-appraisals in adolescents and adults: When social perspective-taking informs self-perception. *Child Development, 80*(4), 1016–1038.

196 **For the few that are profoundly hypnotizable** Crasilneck, H. B., McCranie, E. J., & Jenkins, M. T. (1956). Special indications for hypnosis as a method of anesthesia. *JAMA: Journal of the American Medical Association, 162*(18), 1606–1608; Kosslyn, S. M., Thompson, W. L., Costantini-Ferrando, M. F., Alpert, N. M., & Spiegel, D. (2000). Hypnotic visual illusion alters color processing in the brain. *American Journal of Psychiatry, 157*(8), 1279–1284; Spiegel, H. (1970). A single-treatment method to stop smoking using ancillary self-hypnosis. *International Journal of Clinical and Experimental Hypnosis, 18*(4), 235–250; Surman, O. S., Gottlieb, S. K., Hackett, T. P., & Silverberg, E. L. (1973). Hypnosis in the treatment of warts. *Archives of General Psychiatry, 28*(3), 439.

197 **In our first study, we convinced undergraduates** Falk, E. B., Berkman, E. T., Mann, T., Harrison, B., & Lieberman, M. D. (2010). Predicting persuasion-induced behavior change from the brain. *Journal of Neuroscience, 30,* 8421–8424.

199 **Focus groups don't work all that well** Nisbett, R. E., & Wilson, T. D. (1977). Telling more than we can know: Verbal reports on mental processes. *Psychological Review, 84*(3), 231.

199 **this time we used antismoking ads** Falk, E. B., Berkman, E. T., & Lieberman, M. D. (2011). Neural activity during health messaging predicts reductions in smoking above and beyond self-report. *Health Psychology, 30,* 177–185.

199 **separate the ads based on the advertising campaign** Falk, E. B., Berkman, E. T., & Lieberman, M. D. (2012). From neural responses to population behavior: Neural focus group predicts population level media effects. *Psychological Science, 23,* 439–445.

201 **"Living for others [is] such a relief** Alain de Botton tweet @alaindebotton 3/5/12, 3:00 a.m.

201 **"Only a life lived for others is a life worth while"** *New York Times* (1932). Einstein is terse in rule for success. June 20, p. 17.

201 **I don't really remember what it was like before** Louis C.K. quote from Interview with Jessica Grose, June 17, 2011, in *Slate* titled "Questions for Louis C.K."

202 **Prior to the modern era, humans spent a few years being cared for** Bakan, D. (1971). Adolescence in America: From idea to social fact. *Daedalus, 100,* 979–995; Fasick, F. A. (1994). On the "invention" of adolescence. *Journal of Early Adolescence, 14*(1), 6–23.

202 **Steve Jobs warned the new graduates** Steve Jobs, 2005, Stanford commencement.

Chapter 9: Panoptic Self-Control

203 **Babies have been shown to imitate their parents** Meltzoff, A. N., & Moore, M. K. (1977). Imitation of facial and manual gestures by human neonates. *Science, 198*(4312), 75–78.

203 **babies typically pass the mirror self-recognition test** Amsterdam, B. (1972). Mirror self-image reactions before age two. *Developmental Psychobiology, 5*(4), 297–305.

205 **Mischel tested preschoolers between the ages of three and five** Mischel, W., & Ebbesen, E. B. (1970). Attention in delay of gratification. *Journal of Personality and Social Psychology, 16*(2), 329; Mischel, W., & Baker, N. (1975). Cognitive appraisals and transformations in delay behavior. *Journal of Personality and Social Psychology, 31*(2), 254.

205 **Replacing the actual marshmallows with pictures** Mischel, W., & Moore, B. (1973). Effects of attention to symbolically presented rewards on self-control. *Journal of Personality and Social Psychology, 28*(2), 172.

205 **Mentally focusing on aspects of the marshmallows** Mischel, W., & Baker, N. (1975). Cognitive appraisals and transformations in delay behavior. *Journal of Personality and Social Psychology, 31*(2), 254.

205 **they were able to wait three times as long** Moore, B., Mischel, W., & Zeiss, A. (1976). Comparative effects of the reward stimulus and its cognitive representation in voluntary delay. *Journal of Personality and Social Psychology, 34*(3), 419.

206 **Mischel retested his preschoolers** Shoda, Y., Mischel, W., & Peake, P. K. (1990). Predicting adolescent cognitive and self-regulatory competencies from preschool delay of gratification: Identifying diagnostic conditions. *Developmental Psychology, 26*(6), 978.

206 **preschoolers who could wait** Lehrer, J. (2009). DON'T! The secret of self-control. *New Yorker,* May 18, pp. 26–32.

206 **GPA was better predicted** Duckworth, A. L., & Seligman, M. E. (2005). Self-discipline outdoes IQ in predicting academic performance of adolescents. *Psychological Science, 16*(12), 939–944.

206 **People with higher levels of self-control** Moffitt, T. E., Arseneault, L., Belsky, D., Dickson, N., Hancox, R. J., Harrington, H., . . . , & Caspi, A. (2011). A gradient of childhood self-control predicts health, wealth, and public safety. *Proceedings of the National Academy of Sciences, 108*(7), 2693–2698; Meier, S., & Sprenger, C. D. (2012). Time discounting predicts creditworthiness. *Psychological Science, 23*(1), 56–58; Eisenberg, N., Fabes, R. A., Bernzweig, J., Karbon, M., Poulin, R., & Hanish, L. (1993). The relations of emotionality and regulation to preschoolers' social skills and sociometric status. *Child Development, 64*(5), 1418–1438; Shoda, Y., Mischel, W., & Peake, P. K. (1990). Predicting adolescent cognitive and self-regulatory competencies from preschool

delay of gratification: Identifying diagnostic conditions. *Developmental Psychology*, *26*(6), 978; Tangney, J. P., Baumeister, R. F., & Boone, A. L. (2008). High self-control predicts good adjustment, less pathology, better grades, and interpersonal success. *Journal of Personality*, *72*(2), 271–324; Côté, S., Gyurak, A., & Levenson, R. W. (2010). The ability to regulate emotion is associated with greater well-being, income, and socioeconomic status. *Emotion*, *10*(6), 923.

207 **Our impulses and emotional reactions** Damasio, A. R. (1994). *Descartes' Error*. New York: Putnam.

208 **self-control is like a muscle** Vohs, K. D., & Heatherton, T. F. (2000). Self-regulatory failure: A resource-depletion approach. *Psychological Science*, *11*(3), 249–254; Baumeister, R. F., Bratslavsky, E., Muraven, M., & Tice, D. M. (1998). Ego depletion: Is the active self a limited resource? *Journal of Personality and Social Psychology*, *74*(5), 1252.

208 **It is the only region in the prefrontal cortex** Shaw, P., Lalonde, F., Lepage, C., Rabin, C., Eckstrand, K., Sharp, W., . . . , & Rapoport, J. (2009). Development of cortical asymmetry in typically developing children and its disruption in attention-deficit/hyperactivity disorder. *Archives of General Psychiatry*, *66*(8), 888; Holloway, R. L., & De La Costelareymondie, M. C. (1982). Brain endocast asymmetry in pongids and hominids: Some preliminary findings on the paleontology of cerebral dominance. *American Journal of Physical Anthropology*, *58*(1), 101–110; Zilles, K. (2005). Evolution of the human brain and comparative cyto- and receptor architecture. In S. Dehaene, J. R. Duhamel, M. D. Hauser, & G. Rizzolatti (Eds.). *From Monkey Brain to Human Brain*. Cambridge, MA: MIT Press, Bradford Books, pp. 41–56.

208 **it is appropriate to characterize the rVLPFC** Cohen, J. R., Berkman, E. T., & Lieberman, M. D. (2013). Intentional and incidental self-control in ventrolateral PFC. In D. T. Stuss & R. T. Knight (Eds.). *Principles of Frontal Lobe Function*, 2nd ed. New York: Oxford University Press, pp. 417–440; Cohen, J. R., & Lieberman, M. D. (2010). The common neural basis of exerting self-control in multiple domains. In Y. Trope, R. Hassin, & K. N. Ochsner (Eds.). *Self-control*. New York: Oxford University Press, pp. 141–160.

209 **Countless studies have observed increased rVLPFC** Aron, A. R., Robbins, T. W., & Poldrack, R. A. (2004). Inhibition and the right inferior frontal cortex. *Trends in Cognitive Sciences*, *8*(4), 170–177.

209 **associated with deficits on the no-go task** Aron, A. R., Fletcher, P. C., Bullmore, T., Sahakian, B. J., & Robbins, T. W. (2003). Stop-signal inhibition disrupted by damage to right inferior frontal gyrus in humans. *Nature Neuroscience*, *6*(2), 115–116.

209 **Decades after Mischel performed the initial marshmallow tests** Casey, B. J., Somerville, L. H., Gotlib, I. H., Ayduk, O., Franklin, N. T., Askren,

M. K., . . . , & Shoda, Y. (2011). Behavioral and neural correlates of delay of gratification 40 years later. *Proceedings of the National Academy of Sciences, 108*(36), 14998–15003.

210 **Elliot Berkman and I tested the idea that rVLPFC** Berkman, E. T., Falk, E. B., & Lieberman, M. D. (2011). In the trenches of real-world self-control: Neural correlates of breaking the link between craving and smoking. *Psychological Science, 22,* 498–506.

211 **fewer than half of the participants answer the question correctly** Evans, J. S. B., Barston, J. L., & Pollard, P. (1983). On the conflict between logic and belief in syllogistic reasoning. *Memory & Cognition, 11*(3), 295–306.

211 **Although we may not have as much control** Relative to other kinds of self-control discussed, cognitive self-control is more often lateralized to the left hemisphere rather than to the right. It appears that when cognitive self-control is more wholistic (that is, when one is trying to inhibit an entire thought or belief), it tends to switch back to being right lateralized like the other kinds of self-control described.

211 **To study the neural bases of cognitive self-control** Goel, V., & Dolan, R. J. (2003). Explaining modulation of reasoning by belief. *Cognition, 87*(1), 11–22.

211 **Another study of the belief bias** Tsujii, T., & Watanabe, S. (2010). Neural correlates of belief-bias reasoning under time pressure: A near-infrared spectroscopy study. *NeuroImage, 50*(3), 1320–1326.

212 **a third study used transcranial magnetic stimulation (TMS)** Tsujii, T., Masuda, S., Akiyama, T., & Watanabe, S. (2010). The role of inferior frontal cortex in belief-bias reasoning: An rTMS study. *Neuropsychologia, 48*(7), 2005; Tsujii, T., Sakatani, K., Masuda, S., Akiyama, T., & Watanabe, S. (2011). Evaluating the roles of the inferior frontal gyrus and superior parietal lobule in deductive reasoning: An rTMS study. *NeuroImage, 58*(2), 640–646.

212 **A similar finding has been demonstrated with framing effects** Tversky, A., & Kahneman, D. (1981). The framing of decisions and the psychology of choice. *Science, 211*(4481), 453–458.

212 **An fMRI study examined which brain regions** De Martino, B., Kumaran, D., Seymour, B., & Dolan, R. J. (2006). Frames, biases, and rational decision-making in the human brain. *Science, 313*(5787), 684–687.

213 **Typical fMRI studies of mentalizing** Samson, D., Apperly, I. A., Kathirgamanathan, U., & Humphreys, G. W. (2005). Seeing it my way: A case of a selective deficit in inhibiting self-perspective. *Brain, 128*(5), 1102–1111; van der Meer, L., Groenewold, N. A., Nolen, W. A., Pijnenborg, M., & Aleman, A. (2011). Inhibit yourself and understand the other: Neural basis of distinct processes underlying Theory of Mind. *NeuroImage, 56*(4), 2364–2374.

214 **This is called the *false consensus effect* because we tend** Ross, L., Greene, D., & House, P. (1977). The "false consensus effect": An egocentric bias in social perception and attribution processes. *Journal of Experimental Social Psychology*, *13*(3), 279–301.

216 **suppression isn't used to suppress one's experience of an emotion** Gross, J. J. (2002). Emotion regulation: Affective, cognitive, and social consequences. *Psychophysiology*, *39*(3), 281–291.

217 **"Pain is inevitable** Murakami, H. (2008). *What I Talk About When I Talk About Running: A Memoir*. New York: Knopf, p. vii.

217 **our reality derives from the stories we tell ourselves** Bower, J. E., Low, C. A., Moskowitz, J. T., Sepah, S., & Epel, E. (2007). Benefit finding and physical health: Positive psychological changes and enhanced allostasis. *Social and Personality Psychology Compass*, *2*(1), 223–244.

217 **These threat reactions are orchestrated** Pape, H. C. (2010). Petrified or aroused with fear: The central amygdala takes the lead. *Neuron*, *67*(4), 527–529.

218 **Suppression and reappraisal differ** Butler, E. A., Egloff, B., Wilhelm, F. H., Smith, N. C., Erickson, E. A., & Gross, J. J. (2003). The social consequences of expressive suppression. *Emotion*, *3*(1), 48; Richards, J. M., & Gross, J. J. (2000). Emotion regulation and memory: The cognitive costs of keeping one's cool. *Journal of Personality and Social Psychology*, *79*(3), 410; Gross, J. J. (2002). Emotion regulation: Affective, cognitive, and social consequences. *Psychophysiology*, *39*(3), 281–291.

218 **Despite these differences between suppression and reappraisal** Ochsner, K. N., & Gross, J. J. (2005). The cognitive control of emotion. *Trends in Cognitive Sciences*, *9*(5), 242–249.

218 **For people who reappraise** Goldin, P. R., McRae, K., Ramel, W., & Gross, J. J. (2008). The neural bases of emotion regulation: Reappraisal and suppression of negative emotion. *Biological Psychiatry*, *63*(6), 577.

219 **VLPFC activity is linked to our success** Lee, T.-W., Dolan, R. J., & Critchley, H. D. (2008). Controlling emotional expression: Behavioral and neural correlates of nonimitative emotional responses. *Cerebral Cortex*, *18*(1), 104–113.

219 **In reappraisal, VLPFC activity has been linked** Ochsner, K. N., Bunge, S. A., Gross, J. J., & Gabrieli, J. D. (2002). Rethinking feelings: An fMRI study of the cognitive regulation of emotion. *Journal of Cognitive Neuroscience*, *14*(8), 1215–1229; Phan, K. L., Fitzgerald, D. A., Nathan, P. J., Moore, G. J., Uhde, T. W., & Tancer, M. E. (2005). Neural substrates for voluntary suppression of negative affect: A functional magnetic resonance imaging study. *Biological Psychiatry*, *57*, 210–219; Kalisch, R. (2009). The functional neuroanatomy of reappraisal: Time matters. *Neuroscience & Biobehavioral Reviews*, *33*(8), 1215–1226; Kalisch, R., Wiech, K., Critchley, H. D., Seymour, B., O'Doherty, J. P.,

Oakley, D. A., . . . , & Dolan, R. J. (2005). Anxiety reduction through detachment: Subjective, physiological, and neural effects. *Journal of Cognitive Neuroscience, 17*(6), 874–883.

219 **The longer the time a person spends reappraising** Kalisch, R. (2009). The functional neuroanatomy of reappraisal: Time matters. *Neuroscience and Biobehavioral Reviews, 33,* 1215–1226

219 **Putting our feelings into words** Pennebaker, J. W., & Beall, S. K. (1986). Confronting a traumatic event: Toward an understanding of inhibition and disease. *Journal of Abnormal Psychology, 95,* 274–281.

219 **Preschoolers who can describe their feelings** Denham, S. A. (1986). Social cognition, prosocial behavior, and emotion in preschoolers: Contextual validation. *Child Development, 57,* 194–201; Denham, S. A., & Burton, R. (1996). A social-emotional intervention for at-risk 4-year-olds. *Journal of School Psychology, 34,* 225–245; Fabes, R. A., Eisenberg, N., Hanish, L. D., & Spinrad, T. L. (2001). Preschoolers' spontaneous emotion vocabulary: Relations to likability. *Early Education & Development, 12,* 11–27; Fujiki, M., Brinton, B., & Clarke, D. (2002). Emotion regulation in children with specific language impairment. *Language, Speech, and Hearing Services in Schools, 33,* 102–111; Izard, C., Fine, S., Schultz, D., Mostow, A., Ackerman, B., & Youngstrom, E. (2001). Emotion knowledge as a predictor of social behavior and academic competence in children at risk. *Psychological Science, 12,* 18–23; Mostow, A. J., Izard, C. E., Fine, S., & Trentacosta, C. J. (2002). Modeling emotional, cognitive, and behavioral predictors of peer acceptance. *Child Development, 73,* 1775–1787.

219 **High school students who write about** Ramirez, G., & Beilock, S. L. (2011). Writing about testing worries boosts exam performance in the classroom. *Science, 331,* 211–213.

219 **We have found that labeling the affective aspect** Lieberman, M. D., Inagaki, T. K., Tabibnia, G., & Crockett, M. J. (2011). Subjective responses to emotional stimuli during labeling, reappraisal, and distraction. *Emotion, 3,* 468–480; Burklund, L. J., Creswell, J. D., Irwin, M. R., & Lieberman, M. D. (under review). The common neural bases of affect labeling and reappraisal.

220 **we found that affect labeling helped the most** Kircanski, K., Lieberman, M. D., & Craske, M. G. (2012). Feelings into words: Contributions of language to exposure therapy. *Psychological Science, 23,* 1086–1091.

220 **When people label an emotional picture** Hariri, A. R., Bookheimer, S. Y., & Mazziotta, J. C. (2000). Modulating emotional responses: Effects of a neocortical network on the limbic system. *Neuroreport, 11*(1), 43–48; Lieberman, M. D., Eisenberger, N. I., Crockett, M. J., Tom, S., Pfeifer, J. H., & Way, B. M. (2007). Putting feelings into words: Affect labeling disrupts amygdala activity to affective stimuli. *Psychological Science, 18,*

421–428; Burklund, L. J., Creswell, J. D., Irwin, M. R., & Lieberman, M. D. (under review). The common neural bases of affect labeling and reappraisal; Payer, D. E., Baicy, K., Lieberman, M. D., & London, E. D. (2012). Overlapping neural substrates between intentional and incidental down-regulation of negative emotions. *Emotion*, *12*(2), 229.

221 **We have seen similar things going on in the rVLPFC** Payer, D. E., Baicy, K., Lieberman, M. D., & London, E. D. (2012). Overlapping neural substrates between intentional and incidental down-regulation of negative emotions. *Emotion*, *2*, 229–235; Burklund, L. J., Creswell, J. D., Irwin, M. R., & Lieberman, M. D. (under review). The common neural bases of affect labeling and reappraisal.

223 **George is merely "experience—an entity experiencing"** Isherwood, C. (2001). *A Single Man*. London: Vintage Books, p. 11.

224 **Fewer than half of the doctors in the United States** Adams, S. (2012). Why do so many doctors regret their job choice? *Forbes.com*, April 27: http://www.forbes.com/sites/susanadams/2012/04/27/why-do-so-many-doctors-regret-their-job-choice/.

224 *bang ye,* **which literally means "exposing grandfathers"** Glionna, J. (2010). China tries in vain to keep bellies buttoned up. *Los Angeles Times*, August 10.

225 **Both with strangers and with our romantic partners** Righetti, F., & Finkenauer, C. (2011). If you are able to control yourself, I will trust you: The role of perceived self-control in interpersonal trust. *Journal of Personality and Social Psychology*, *100*(5), 874.

225 **This makes good sense in the case of romantic partners** Pronk, T. M., Karremans, J. C., & Wigboldus, D. H. (2011). How can you resist? Executive control helps romantically involved individuals to stay faithful. *Journal of Personality and Social Psychology*, *100*(5), 827.

225 **the creators of the SAT designed it to be a measure of intelligence** Gladwell, M. (2001, December 17). Examined life: What Stanley H. Kaplan taught us about the SAT. *New Yorker*, 86.

226 **"socialized behavior is thus the supreme achievement** Allport, F. H. (1924). *Social Psychology*. Boston: Houghton Mifflin Company, p. 31.

227 **"I have a lot of beliefs, and I live by none of them"** Louis C.K. Live at the Beacon Theater.

227 **"morals reformed—health preserved—industry invigorated"** Bentham, J. (1995). *The Panopticon Writings*. Edited by M. Bozovic. London: Verso, pp. 29–95.

229 **The subject was 30 percent more likely** van Rompay, T. J., Vonk, D. J., & Fransen, M. L. (2009). The eye of the camera effects of security cameras on prosocial behavior. *Environment and Behavior*, *41*(1), 60–74.

229 **Another study found that individuals were twice as likely to cheat** Zhong, C. B., Bohns, V. K., & Gino, F. (2010). Good lamps are the

best police: Darkness increases dishonesty and self-interested behavior. *Psychological Science, 21*(3), 311–314.

229 **subjects wearing eye tracking devices** Risko, E. F., & Kingstone, A. (2011). Eyes wide shut: Implied social presence, eye tracking and attention. *Attention, Perception, & Psychophysics, 73*(2), 291–296.

229 **led people to pay 276 percent more into the honesty box** Bateson, M., Nettle, D., & Roberts, G. (2006). Cues of being watched enhance cooperation in a real-world setting. *Biology Letters, 2*(3), 412–414.

229 **a similar "eyes poster" nearly halved the amount of littering** Ernest-Jones, M., Nettle, D., & Bateson, M. (2011). Effects of eye images on everyday cooperative behavior: A field experiment. *Evolution and Human Behavior, 32*(3), 172–178; see also Powell, K. L., Roberts, G., & Nettle, D. (2012). Eye images increase charitable donations: Evidence from an opportunistic field experiment in a supermarket. *Ethology, 118,* 1096–1101; Nettle, D., Nott, K., & Bateson, M. (2012). "Cycle thieves, we are watching you": Impact of a simple signage intervention against bicycle theft. *PLOS One, 7*(12), e51738.

229 **Even pictures of a defunct toy robot's eyes** Burnham, T. C., & Hare, B. (2007). Engineering human cooperation. *Human Nature, 18*(2), 88–108.

230 **Men presented with the "face" version of the three dots** Rigdon, M., Ishii, K., Watabe, M., & Kitayama, S. (2009). Minimal social cues in the dictator game. *Journal of Economic Psychology, 30*(3), 358–367.

231 **When children (ages nine and up) were put in this scenario** Beaman, A. L., Klentz, B., Diener, E., & Svanum, S. (1979). Self-awareness and transgression in children: Two field studies. *Journal of Personality and Social Psychology, 37*(10), 1835.

231 **self-consciousness is essentially a dialogue** Mead, G. H. (1934). *Mind, Self, and Society from the Standpoint of a Social Behaviorist.* Edited by C. W. Morris. Chicago: University of Chicago; Cooley, C. H. (1902). *Human Nature and the Social Order.* New York: Scribner.

231 **first-year college students were ten times less likely** Diener, E., & Wallbom, M. (1976). Effects of self-awareness on antinormative behavior. *Journal of Research in Personality, 10*(1), 107–111.

231 **People are also more likely to conform** Abrams, D., & Brown, R. (1989). Self-consciousness and social identity: Self-regulation as a group member. *Social Psychology Quarterly, 52,* 311–318; Duval, S. (1976). Conformity on a visual task as a function of personal novelty on attitudinal dimensions and being reminded of the object status of self. *Journal of Experimental Social Psychology, 12*(1), 87–98; Swart, C., Ickes, W., & Morgenthaler, E. S. (1978). The effect of objective self awareness on compliance in a reactance situation. *Social Behavior and Personality: An International Journal, 6*(1), 135–139.

233 **Manfred Spitzer and Ernst Fehr ran this study** Spitzer, M., Fischbacher, U., Herrnberger, B., Grön, G., & Fehr, E. (2007). The neural signature of social norm compliance. *Neuron, 56*(1), 185–196.

234 **People who conform most to this kind of norm** Campbell-Meiklejohn, D. K., Bach, D. R., Roepstorff, A., Dolan, R. J., & Frith, C. D. (2010). How the opinion of others affects our valuation of objects. *Current Biology, 20*(13), 1165–1170; Campbell-Meiklejohn, D. K., Kanai, R., Bahrami, B., Bach, D. R., Dolan, R. J., Roepstorff, A., & Frith, C. D. (2012). Structure of orbitofrontal cortex predicts social influence. *Current Biology, 22*(4), R123–R124.

234 **there is research indicating that just imagining what others think of you** Pfeifer, J. H., Masten, C. L., Borofsky, L. A., Dapretto, M., Fuligni, A. J., & Lieberman, M. D. (2009). Neural correlates of direct and reflected self-appraisals in adolescents and adults: When social perspective-taking informs self-perception. *Child Development, 80*(4), 1016–1038; Ochsner, K. N., Beer, J. S., Robertson, E. R., Cooper, J. C., Gabrieli, J. D., Kihsltrom, J. F., & D'Esposito, M. (2005). The neural correlates of direct and reflected self-knowledge. *NeuroImage, 28*(4), 797–814.

234 **Can you guess which brain region** Lieberman, M. D. (2007). Social cognitive neuroscience: A review of core processes. *Annual Review of Psychology, 58*, 259–289.

Chapter 10: Living with a Social Brain

242 **In 1989, more than 200,000 college freshmen were asked** Easterlin, R. A., & Crimmins, E. M. (1991). Private materialism, personal self-fulfillment, family life, and public interest: The nature, effects, and causes of recent changes in the values of American youth. *Public Opinion Quarterly, 55*(4), 499–533.

243 **Economists have been obsessed with this question** Easterlin, R. A. (1974). Does economic growth improve the human lot? In P. A. David and M. W. Reder (Eds.). *Nations and Households in Economic Growth: Essays in Honour of Moses Abramovitz.* New York: Academic Press; Diener, E., & Seligman, M. E. (2004). Beyond money. *Psychological Science in the Public Interest, 5*(1), 1–31.

243 **If we look at a large number of countries** Diener, E., Sandvik, E., Seidlitz, L., & Diener, M. (1993). The relationship between income and subjective well-being: Relative or absolute? *Social Indicators Research, 28*(3), 195–223.

244 **happiness researcher Ed Diener looked at surveys of thousands of U.S. adults** Ibid.

244 **one study examined changing U.S. income levels between 1946 and**

1990 Diener, E., & Suh, E. (1997). Measuring quality of life: Economic, social, and subjective indicators. *Social Indicators Research, 40,* 189–216.

244 **This effect, called the *Easterlin Paradox*** Easterlin, R. A. (1974). Does economic growth improve the human lot? In P. A. David and M. W. Reder (Eds.). *Nations and Households in Economic Growth: Essays in Honour of Moses Abramovitz.* New York: Academic Press; Diener, E., & Seligman, M. E. (2004). Beyond money. *Psychological Science in the Public Interest, 5*(1), 1–31.

244 **real income increased 500 percent** Easterlin, R. A. (1995). Will raising the incomes of all increase the happiness of all? *Journal of Economic Behavior & Organization, 27*(1), 35–47.

246 **Psychologists pointed out that humans** Frederick, S., & Loewenstein, G. (1999). Hedonic adaptation. In D. Kanheman & E. Diener (Eds.). *The Foundations of Hedonic Psychology.* New York: Russell Sage Foundation, pp. 302–329.

246 **They reported being no happier** Brickman, P., Coates, D., & Janoff-Bulman, R. (1978). Lottery winners and accident victims: Is happiness relative? *Journal of Personality and Social Psychology, 36*(8), 917.

246 **This *relative income* argument suggests** Kahneman, D., Krueger, A. B., Schkade, D., Schwarz, N., & Stone, A. A. (2006). Would you be happier if you were richer? A focusing illusion. *Science, 312*(5782), 1908–1910.

246 **In the book *Bowling Alone* Robert Putman** Putnam, R. D. (2000). *Bowling Alone: The Collapse and Revival of American Community.* New York: Simon & Schuster.

247 **Pretty much any way economists examine** Ibid.; Helliwell, J. F., & Putnam, R. D. (2004). The social context of well-being. *Philosophical Transactions of the Royal Society of London Series B: Biological Sciences,* 1435–1446.

247 **One study compared the impact of income** Becchetti, L., Pelloni, A., & Rossetti, F. (2008). Relational goods, sociability, and happiness. *Kyklos, 61*(3), 343–363.

247 **volunteering was associated with greater well-being** Borgonovi, F. (2008). Doing well by doing good: The relationship between formal volunteering and self-reported health and happiness. *Social Science & Medicine, 66*(11), 2321–2334.

247 **giving to charity is related to changes in well-being** Aknin, L. B., Barrington-Leigh, C. P., Dunn, E. W., Helliwell, J. F., Biswas-Diener, R., Kemeza, I., . . . , & Norton, M. I. (2010). Prosocial spending and well-being: Cross-cultural evidence for a psychological universal (No. w16415). National Bureau of Economic Research.

247 **having a friend whom you see on most days,** Powdthavee, N. (2008). Putting a price tag on friends, relatives, and neighbours: Using surveys

of life satisfaction to value social relationships. *Journal of Socio-economics, 37*(4), 1459–1480.

248 **social factors are also huge determinants of physical health** Holt-Lunstad, J., Smith, T. B., & Layton, J. B. (2010). Social relationships and mortality risk: A meta-analytic review. *PLOS Medicine, 7*(7), e1000316.

248 **People are significantly less likely to be married today** Bumpass, L. L., Sweet, J. A., & Cherlin, A. (1991). The role of cohabitation in declining rates of marriage. *Journal of Marriage and the Family, 53,* 913–927; Popenoe, D. (1993). American family decline, 1960–1990: A review and appraisal. *Journal of Marriage and the Family,* 55, 527–542.

248 **We volunteer less, participate in fewer social groups** Costa, D. L., & Kahn, M. E. (2001). Understanding the decline in social capital, 1952–1998 (No. w8295). National Bureau of Economic Research; Putnam, R. D. (2000). *Bowling Alone: The Collapse and Revival of American Community.* New York: Simon & Schuster.

248 **people were asked to list their friends** McPherson, M., Smith-Lovin, L., & Brashears, M. E. (2006). Social isolation in America: Changes in core discussion networks over two decades. *American Sociological Review, 71*(3), 353–375.

250 **In 1965, only 45 percent of college freshmen** Easterlin, R. A., & Crimmins, E. M. (1991). Private materialism, personal self-fulfillment, family life, and public interest: The nature, effects, and causes of recent changes in the values of American youth. *Public Opinion Quarterly, 55*(4), 499–533.

250 **more individuals endorse materialism** Nickerson, C., Schwarz, N., Diener, E., & Kahneman, D. (2003). Zeroing in on the dark side of the American dream: A closer look at the negative consequences of the goal for financial success. *Psychological Science, 14*(6), 531–536; Chan, R., & Joseph, S. (2000). Dimensions of personality, domains of aspiration, and subjective well-being. *Personality and Individual Differences, 28*(2), 347–354.

252 **Approximately a third of all Americans live in apartments** NMHC tabulations of 2012 Current Population Survey, Annual Social and Economic Supplement, U.S. Census Bureau (http://www.census.gov /cps). Updated October 2012.
http://www.nmhc.org/Content.cfm?ItemNumber=55508.

254 **when people are prompted to think about time** Mogilner, C. (2010). The pursuit of happiness. *Psychological Science, 21*(9), 1348–1354.

254 **benefits of socializing through social snacking** Gardner, W. L., Pickett, C. L., & Knowles, M. (2005). Social snacking and shielding. In K. D. Williams, J. P. Forgas, & W. V. Hippel (Eds.). *The Social Outcast: Ostracism, Social Exclusion, Rejection, & Bullying.* New York: Psychology Press.

254 **they reported the pain to be less painful** Master, S. L., Eisenberger, N. I., Taylor, S. E., Naliboff, B. D., & Lieberman, M. D. (2009). A picture's worth: Partner photographs reduce experimentally induced pain. *Psychological Science, 20,* 1316–1318; Eisenberger, N. I., Master, S. L., Inagaki, T. K., Taylor, S. E., Shirinyan, D., Lieberman, M. D., & Naliboff, B. (2011). Attachment figures activate a safety signal-related neural region and reduce pain experience. *Proceedings of the National Academy of Sciences, 108,* 11721–11726.

254 **digital picture frames to people in hospitals** http://www.nikon-kraftderbilder.de.

255 **we are more motivated to tune in to our favorite shows** Derrick, J. L., Gabriel, S., & Hugenberg, K. (2009). Social surrogacy: How favored television programs provide the experience of belonging. *Journal of Experimental Social Psychology, 45*(2), 352–362.

255 **it is also likely to "crowd out"** Bruni, L., & Stanca, L. (2008). Watching alone: Relational goods, television and happiness. *Journal of Economic Behavior & Organization, 65*(3), 506–528.

255 **first seminal study examining these questions** Kraut, R., Patterson, M., Lundmark, V., Kiesler, S., Mukophadhyay, T., & Scherlis, W. (1998). Internet paradox: A social technology that reduces social involvement and psychological well-being? *American Psychologist, 53*(9), 1017.

255 **A series of other papers came out soon after** Valkenburg, P. M., & Peter, J. (2009). Social consequences of the Internet for adolescents: A decade of research. *Current Directions in Psychological Science, 18*(1), 1–5.

256 **Because Facebook use is more of an extension** Ellison, N. B., Steinfield, C., & Lampe, C. (2007). The benefits of Facebook "friends": Social capital and college students' use of online social network sites. *Journal of Computer-Mediated Communication, 12*(4), 1143–1168; Grieve, R., Indian, M., Witteveen, K., Anne Tolan, G., & Marrington, J. (2013). Face-to-face or Facebook: Can social connectedness be derived online? *Computers in Human Behavior, 29*(3), 604–609; Steinfield, C., Ellison, N. B., & Lampe, C. (2008). Social capital, self-esteem, and use of online social network sites: A longitudinal analysis. *Journal of Applied Developmental Psychology, 29*(6), 434–445.

Chapter 11: The Business of Social Brains

258 **"Economists presume that [people] do not work** Camerer, C. F., & Hogarth, R. M. (1999). The effects of financial incentives in experiments: A review and capital-labor-production framework. *Journal of Risk and Uncertainty, 19*(1), 7–42.

258 **pay for performance usually produces** Ibid.; Jenkins Jr., G. D., Mitra, A., Gupta, N., & Shaw, J. D. (1998). Are financial incentives related to

performance? A meta-analytic review of empirical research. *Journal of Applied Psychology, 83*(5), 777.

259 **Rock has developed the SCARF model** Rock, D. (2009). Managing with the brain in mind. *Strategy + Business, 56*, 58–67.

259 **"the primary colors of intrinsic motivation"** Bryant, A. (2013). A boss's challenge: Have everyone join the "in" group. *New York Times,* March 23.

259 **Autonomy and certainty aren't really a social part of the story** Pink, D. H. (2010). *Drive: The Surprising Truth About What Motivates Us.* New York: Canongate.

260 **how much people crave status and recognition** Larkin, I. (2010). Paying $30,000 for a gold star: An empirical investigation into the value of peer recognition to software salespeople. Working paper, Harvard Business School, Boston.

262 **companies with more human capital** Bourdieu, P. (1986). The forms of capital. In J. G. Richardson (Ed.). *Handbook of Theory and Research for the Sociology of Education.* New York: Greenwood, pp. 241–258; Putnam, R. D. (2000). *Bowling Alone: The Collapse and Revival of American Community.* New York: Simon & Schuster.

262 **Economist Arent Greve studied three Italian** Greve, A., Benassi, M., & Sti, A. D. (2010). Exploring the contributions of human and social capital to productivity. *International Review of Sociology—Revue Internationale de Sociologie, 20*(1), 35–58.

263 **Social connections are essentially the original Internet** Bosma, N., Van Praag, M., Thurik, R., & De Wit, G. (2004). The value of human and social capital investments for the business performance of startups. *Small Business Economics, 23*(3), 227–236; Chen, M. H., Chang, Y. C., & Hung, S. C. (2007). Social capital and creativity in R&D project teams. *R&D Management, 38*(1), 21–34.

263 **The extent to which employees perceive decisions** Colquitt, J. A., Conlon, D. E., Wesson, M. J., Porter, C. O., & Ng, K. Y. (2001). Justice at the millennium: A meta-analytic review of 25 years of organizational justice research. *Journal of Applied Psychology, 86*(3), 425.

263 **Fairness might seem like a squishy motivator** Tabibnia, G., Satpute, A. B., & Lieberman, M. D. (2008). The sunny side of fairness: Preference for fairness activates reward circuitry (and disregarding unfairness activates self-control circuitry). *Psychological Science, 19,* 339–347.

264 **the chance to help others motivates people** Grant, A. M. (2013). *Give and Take: A Revolutionary Approach to Success.* New York: Viking.

264 **he focused on people working at a university** Grant, A. M., Campbell, E. M., Chen, G., Cottone, K., Lapedis, D., & Lee, K. (2007). Impact and the art of motivation maintenance: The effects of contact with beneficiaries on persistence behavior. *Organizational Behavior and Human Decision Processes, 103*(1), 53–67.

265 **Grant replaced the face-to-face meetings** Grant, A. M. (2008). The significance of task significance: Job performance effects, relational mechanisms, and boundary conditions. *Journal of Applied Psychology, 93*(1), 108.

266 **Grant's second approach to caring and workplace** Grant, A. M., Dutton, J. E., & Rosso, B. D. (2008). Giving commitment: Employee support programs and the prosocial sensemaking process. *Academy of Management Journal, 51*(5), 898–918.

266 **Although he did not measure job performance directly** Harter, J. K., Schmidt, F. L., & Hayes, T. L. (2002). Business-unit-level relationship between employee satisfaction, employee engagement, and business outcomes: A meta-analysis. *Journal of Applied Psychology, 87*(2), 268.

266 **"He that has once done you a kindness** Franklin, B. (1868/1996). *The Autobiography of Benjamin Franklin.* New York: Dover, p. 80.

267 **When we see ourselves doing something** Bem, D. J. (1972). Self-perception theory. In L. Berkowitz (Ed.). *Advances in Experimental Social Psychology* (Vol. 6). New York: Academic Press, pp. 1– 62; Burger, J. M. (1999). The foot-in-the-door compliance procedure: A multiple-process analysis and review. *Personality and Social Psychology Review, 3*(4), 303–325.

268 **they would prefer a better boss to a higher salary** National Boss Day Poll (America 2012) (www.tellyourboss.com).

268 **Some managers might feel that being disliked** Ibid.

269 **asked thousands of employees to score the leadership** Zenger, J., & Folkman, J. (2009). *The Extraordinary Leader: Turning Good Managers into Great Leaders.* New York: McGraw-Hill.

270 **three-person teams were brought together** Kellett, J. B., Humphrey, R. H., & Sleeth, R. G. (2006). Empathy and the emergence of task and relations leaders. *Leadership Quarterly, 17*(2), 146–162.

271 **the characteristics that people associated with leaders** Lord, R. G., De Vader, C. L., & Alliger, G. M. (1986). A meta-analysis of the relation between personality traits and leadership perceptions: An application of validity generalization procedures. *Journal of Applied Psychology, 71*(3), 402.

272 **One study examined this possibility by looking at the relationships** Kellett, J. B., Humphrey, R. H., & Sleeth, R. G. (2002). Empathy and complex task performance: Two routes to leadership. *Leadership Quarterly, 13*(5), 523–544.

272 **Though there have been studies showing the social and nonsocial reasoning** Meyer, M. L., Spunt, R. P., Berkman, E. T., Taylor, S. E., & Lieberman, M. D. (2012). Social working memory: An fMRI study of parametric increases in social cognitive effort. *Proceedings of the National Academy of Sciences, 109*, 1883–1888; Spreng, N., Stevens, W. D., Chamberlain, J. P., Gilmore, A. W., and Schacter, D. L. (2010). Default

network activity, coupled with the frontoparietal control network, supports goal-directed cognition. *NeuroImage 53,* 303–317; Christoff, K., Gordon, A. M., Smallwood, J., Smith, R., and Schooler, J. W. (2009). Experience sampling during fMRI reveals default network and executive system contributions to mind wandering. *Proceedings of the National Academy of Sciences of the United States of America, 106,* 8719–8724.

Chapter 12: Educating the Social Brain

275 **In the United States, we spend more on public education (kindergarten through twelfth grade)** http://www.usgovernmentspending.com/us_education_spending_20.html.

275 **Out of 34 comparison countries** OECD Programme for International Student Assessment (PISA) (2009): http://www.oecd.org/pisa/pisaproducts/pisa2009/pisa2009keyfindings.htm. Executive Summary: http://www.oecd.org/pisa/pisaproducts/46619703.pdf.

275 **we are getting a lousy return** Even just within the United States, increases in education spending in each state over the past two decades have had very little connection to the achievement gains seen in those states. Hanushek, E. A., Peterson, P. E., & Woessmann, L. (2012). Is the U.S. catching up? International and state trends in student achievement. *Education Next,* 24–33.

275 **the societal payoff would be immeasurable** Juvonen, J., et al. (2004). *Focus on the Wonder Years: Challenges Facing the American Middle School* (Vol. 139). Santa Monica, CA: RAND Corporation; Eccles, J. S., Midgley, C., Wigfield, A., Buchanan, C. M., Reuman, D., Flanagan, C., & Mac Iver, D. (1993). Development during adolescence: The impact of stage-environment fit on young adolescents' experiences in schools and in families. *American Psychologist, 48*(2), 90.

276 **There are myriad reasons why academic performance** Baumeister, R. F., & Leary, M. R. (1995). The need to belong: Desire for interpersonal attachments as a fundamental human motivation. *Psychological Bulletin, 117*(3), 497.

276 **This switch also brings with it a change** Eccles, J. S., Midgley, C., Wigfield, A., Buchanan, C. M., Reuman, D., Flanagan, C., & Mac Iver, D. (1993). Development during adolescence: The impact of stage-environment fit on young adolescents' experiences in schools and in families. *American Psychologist, 48*(2), 90.

277 **Do junior high students feel like they don't belong?** Juvonen, J. (2004). *Focus on the Wonder Years: Challenges Facing the American Middle School* (Vol. 139). Santa Monica, CA: RAND Corporation.

277 **Observers' inaction is taken as a tacit endorsement** Juvonen, J., & Galván, A. (2009). Bullying as a means to foster compliance. In M.

Harris (Ed). *Bullying, Rejection and Peer Victimization: A Social Cognitive Neuroscience Perspective.* New York: Springer, pp. 299–318.

277 **bullying in school is associated with negative changes** Fekkes, M., Pijpers, F. I., Fredriks, A. M., Vogels, T., & Verloove-Vanhorick, S. P. (2006). Do bullied children get ill, or do ill children get bullied? A prospective cohort study on the relationship between bullying and health-related symptoms. *Pediatrics, 117*(5), 1568–1574; Nishina, A., Juvonen, J., & Witkow, M. R. (2005). Sticks and stones may break my bones, but names will make me feel sick: The psychosocial, somatic, and scholastic consequences of peer harassment. *Journal of Clinical Child and Adolescent Psychology, 34*(1), 37–48.

278 **Young adolescents experiencing more bullying** Juvonen, J., Nishina, A., & Graham, S. (2000). Peer harassment, psychological adjustment, and school functioning in early adolescence. *Journal of Educational Psychology, 92*(2), 349; Lopez, C., & DuBois, D. L. (2005). Peer victimization and rejection: Investigation of an integrative model of effects on emotional, behavioral, and academic adjustment in early adolescence. *Journal of Clinical Child and Adolescent Psychology, 34*(1), 25–36.

278 **as many as 40 percent of adolescents report** Wang, J., Iannotti, R. J., Luk, J. W., & Nansel, T. R. (2010). Co-occurrence of victimization from five subtypes of bullying: Physical, verbal, social exclusion, spreading rumors, and cyber. *Journal of Pediatric Psychology, 35*(10), 1103–1112.

278 **Dewey Cornell examined how schools as a whole** Lacey, A., & Cornell, D. (under review). The impact of teasing and bullying on school-wide academic performance.

278 **chronic physical pain is associated with** Dick, B. D., & Rashiq, S. (2007). Disruption of attention and working memory traces in individuals with chronic pain. *Anesthesia & Analgesia, 104*(5), 1223–1229; Glass, J. M. (2009). Review of cognitive dysfunction in fibromyalgia: A convergence on working memory and attentional control impairments. *Rheumatic Disease Clinics of North America, 35,* 299–311.

278 **social pain leads to decrements in intellectual performance** Baumeister, R. F., Twenge, J. M., & Nuss, C. K. (2002). Effects of social exclusion on cognitive processes: Anticipated aloneness reduces intelligent thought. *Journal of Personality and Social Psychology, 83*(4), 817.

279 **a modest impact on GPA of being accepted** Chen, X., Rubin, K. H., & Li, D. (1997). Relation between academic achievement and social adjustment: Evidence from Chinese children. *Developmental Psychology, 33*(3), 518; Furrer, C., & Skinner, E. (2003). Sense of relatedness as a factor in children's academic engagement and performance. *Journal of Educational Psychology, 95*(1), 148; Wentzel, K. R., & Caldwell, K. (1997). Friendships, peer acceptance, and group membership: Relations to academic achievement in middle school. *Child Development, 68*(6),

1198–1209; Wentzel, K. R. (1998). Social relationships and motivation in middle school: The role of parents, teachers, and peers. *Journal of Educational Psychology, 90*(2), 202.

279 **The most persuasive findings** Walton, G. M., & Cohen, G. L. (2007). A question of belonging: Race, social fit, and achievement. *Journal of Personality and Social Psychology, 92*(1), 82; Walton, G. M., & Cohen, G. L. (2011). A brief social-belonging intervention improves academic and health outcomes of minority students. *Science, 331*(6023), 1447–1451.

279 **they tested the effects on African-American** http://oir.yale.edu/yale-factsheet.

280 **Alice Isen repeatedly observed that feeling good** Isen, A. M., Daubman, K. A., & Nowicki, G. P. (1987). Positive affect facilitates creative problem solving. *Journal of Personality and Social Psychology, 52*(6), 1122.

280 **positive affect enhances working memory ability** Carpenter, S. M., Peters, E., Västfjäll, D., & Isen, A. M. (2013). Positive feelings facilitate working memory and complex decision making among older adults. *Cognition and Emotion, 27*, 184–192; Esmaeili, M. T., Karimi, M., Tabatabaie, K. R., Moradi, A., & Farahini, N. (2011). The effect of positive arousal on working memory. *Procedia: Social and Behavioral Sciences, 30*, 1457–1460.

280 **feeling good and thinking well both depend on dopamine** Ashby, F. G., & Isen, A. M. (1999). A neuropsychological theory of positive affect and its influence on cognition. *Psychological Review, 106*(3), 529.

281 **The lateral prefrontal cortex is also rich with dopamine** Aalto, S., Brück, A., Laine, M., Någren, K., & Rinne, J. O. (2005). Frontal and temporal dopamine release during working memory and attention tasks in healthy humans: A positron emission tomography study using the high-affinity dopamine D2 receptor ligand [11C] FLB 457. *Journal of Neuroscience, 25*(10), 2471–2477.

281 **Dopamine reductions in the prefrontal cortex** Brozoski, T. J., Brown, R. M., Rosvold, H. E., & Goldman, P. S. (1979). Cognitive deficit caused by regional depletion of dopamine in prefrontal cortex of rhesus monkey. *Science, 205*, 929–932; Sawaguchi, T., & Goldman-Rakic, P. S. (1991). D1 dopamine receptors in prefrontal cortex: Involvement in working memory. *Science, 251*(4996), 947; Luciana, M., Depue, R. A., Arbisi, P., & Leon, A. (1992). Facilitation of working memory in humans by a D2 dopamine receptor agonist. *Journal of Cognitive Neuroscience, 4*(1), 58–68; Müller, U., Von Cramon, D. Y., & Pollmann, S. (1998). D1- versus D2-receptor modulation of visuospatial working memory in humans. *Journal of Neuroscience, 18*(7), 2720–2728.

281 **If our schools are broken** Compayre, G., & Payne, W. H. (2003). *History of Pedagogy*. New York: Kessinger.

282 **"Thou didst beat me** Longstreet, W. S., & Shane, H. G. (1993). *Curriculum for a New Millennium.* Boston: Allyn & Bacon.

282 **By junior high, education is a battle** Crone, E. A., & Dahl, R. E. (2012). Understanding adolescence as a period of social-affective engagement and goal flexibility. *Nature Reviews Neuroscience, 13*(9), 636–650; Nelson, E. E., Leibenluft, E., McClure, E., & Pine, D. S. (2005). The social re-orientation of adolescence: A neuroscience perspective on the process and its relation to psychopathology. *Psychological Medicine, 35*(02), 163–174; Steinberg, L., & Morris, A. S. (2001). Adolescent development. *Journal of Cognitive Education and Psychology, 2*(1), 55–87.

282 **The mentalizing system that promotes this** Pfeifer, J. H., & Allen, N. B. (2012). Arrested development? Reconsidering dual-systems models of brain function in adolescence and disorders. *Trends in Cognitive Sciences, 16,* 322–329; Blakemore, S. J. (2008). The social brain in adolescence. *Nature Reviews Neuroscience, 9*(4), 267–277.

282 **We spend more than 20,000 hours in classrooms** Conway, M. A., Cohen, G., & Stanhope, N. (1991). On the very long-term retention of knowledge acquired through formal education: Twelve years of cognitive psychology. *Journal of Experimental Psychology: General, 120,* 395–409.

283 **We need the social brain to work for us,** There are some great start-ups trying to do exactly this. See Rob Hutter's *Learn Capital:* http://www.learncapital.com.

283 **Classroom learning as it typically occurs** Wagner, A. D., Schacter, D. L., Rotte, M., Koutstaal, W., Maril, A., Dale, A. M., . . . , & Buckner, R. L. (1998). Building memories: Remembering and forgetting of verbal experiences as predicted by brain activity. *Science, 281*(5380), 1188–1191.

284 **a series of behavioral studies demonstrated** Hamilton, D. L., Katz, L. B., & Leirer, V. O. (1980). Cognitive representation of personality impressions: Organizational processes in first impression formation. *Journal of Personality and Social Psychology, 39*(6), 1050.

284 **Jason Mitchell, a social neuroscientist at Harvard University** Mitchell, J. P., Macrae, C. N., & Banaji, M. R. (2004). Encoding-specific effects of social cognition on the neural correlates of subsequent memory. *Journal of Neuroscience, 24*(21), 4912–4917.

288 **Yale psychologist John Bargh** Bargh, J. A., & Schul, Y. (1980). On the cog-nitive benefits of teaching. *Journal of Educational Psychology, 72*(5), 593.

290 **multiple studies have demonstrated that peer tutoring benefits** Allen, V. L., & Feldman, R. S. (1973). Learning through tutoring: Low-achieving children as tutors. *Journal of Experimental Education, 42,* 1–5; Rohrbeck, C. A., Ginsburg-Block, M. D., Fantuzzo, J. W., & Miller, T. R. (2003). Peer-assisted learning interventions with elementary school students: A meta-analytic review. *Journal of Educational Psychology, 95*(2),

240; Semb, G. B., Ellis, J. A., & Araujo, J. (1993). Long-term memory for knowledge learned in school. *Journal of Educational Psychology, 85*(2), 305.

292 **Neural and hormonal changes** Nelson, E. E., Leibenluft, E., McClure, E., & Pine, D. S. (2005). The social re-orientation of adolescence: A neuroscience perspective on the process and its relation to psychopathology. *Psychological Medicine, 35*(02), 163–174.

292 **We rarely get direct feedback about our errors** Tesser, A., Rosen, S., & Batchelor, T. R. (1972). On the reluctance to communicate bad news (the MUM effect): A role play extension. *Journal of Personality, 40*(1), 88–103.

294 **The determinist view was buffeted** Rakic, P. (1985). Limits of neurogenesis in primates. *Science, 227*(4690), 1054–1056.

294 **new neurons can be born in adulthood** Gould, E., Reeves, A. J., Graziano, M. S., & Gross, C. G. (1999). Neurogenesis in the neocortex of adult primates. *Science, 286*(5439), 548–552; Buonomano, D. V., & Merzenich, M. M. (1998). Cortical plasticity: From synapses to maps. *Annual Review of Neuroscience, 21*(1), 149–186.

294 **People who were learning to juggle** Draganski, B., Gaser, C., Busch, V., Schuierer, G., Bogdahn, U., & May, A. (2004). Neuroplasticity: Changes in grey matter induced by training. *Nature, 427*(6972), 311–312.

294 **Similarly, taxi drivers in London** Maguire, E. A., Gadian, D. G., Johnsrude, I. S., Good, C. D., Ashburner, J., Frackowiak, R. S., & Frith, C. D. (2000). Navigation-related structural change in the hippocampi of taxi drivers. *Proceedings of the National Academy of Sciences, 97*(8), 4398–4403.

295 **Although working memory capacity and fluid intelligence** Sternberg, R. J. (2008). Increasing fluid intelligence is possible after all. *Proceedings of the National Academy of Sciences, 105*(19), 6791–6792; Jaeggi, S. M., Buschkuehl, M., Jonides, J., & Perrig, W. J. (2008). Improving fluid intelligence with training on working memory. *Proceedings of the National Academy of Sciences, 105*(19), 6829–6833; Buschkuehl, M., Jaeggi, S. M., & Jonides, J. (2012). Neuronal effects following working memory training. *Developmental Cognitive Neuroscience, 25,* S167–S179.

295 **emotionality peaks right around eighth grade** Silvers, J. A., McRae, K., Gabrieli, J. D., Gross, J. J., Remy, K. A., & Ochsner, K. N. (2012). Age-related differences in emotional reactivity, regulation, and rejection sensitivity in adolescence. *Emotion, 12,* 1235–1247; Galvan, A., Hare, T. A., Parra, C. E., Penn, J., Voss, H., Glover, G., & Casey, B. J. (2006). Earlier development of the accumbens relative to orbitofrontal cortex might underlie risk-taking behavior in adolescents. *Journal of Neuroscience, 26*(25), 6885–6892.

296 **From delay of gratification and emotion regulation** Cohen, J. R., Berkman, E. T., & Lieberman, M. D. (2013). Intentional and incidental

self-control in ventrolateral PFC. In D. T. Stuss & R. T. Knight (Eds.). *Principles of Frontal Lobe Function*, 2nd ed. New York: Oxford University Press, pp. 417–440.

296 **examined the effects of training motor self-control** Morales, J. I., Berkman, E. T., & Lieberman, M. D. (2012). Improving self-control across domains: Increasing emotion regulation ability through motor inhibition training. Unpublished manuscript; Muraven, M. (2010). Building self-control strength: Practicing self-control leads to improved self-control performance. *Journal of Experimental Social Psychology, 46,* 465–468; Muraven, M. (2010). Practicing self-control lowers the risk of smoking lapse. *Psychology of Addictive Behaviors, 24*(3), 446; Schweizer, S., Grahn, J., Hampshire, A., Mobbs, D., & Dalgleish, T. (2013). Training the emotional brain: Improving affective control through emotional working memory training. *Journal of Neuroscience, 33,* 5301–5311

297 **mindfulness meditation may turn out to be a great way** Creswell, J. D., Burklund, L. J., Irwin, M. R., & Lieberman, M. D. (in prep). Mindfulness meditation training increases functional activity in right ventrolateral prefrontal cortex during affect labeling in older adults: A randomized controlled study; Farb, N. A., Segal, Z. V., Mayberg, H., Bean, J., McKeon, D., Fatima, Z., & Anderson, A. K. (2007). Attending to the present: Mindfulness meditation reveals distinct neural modes of self-reference. *Social Cognitive and Affective Neuroscience, 2*(4), 313–322.

Epilogue

299 **It is useless to attempt to reason a man** Ballou, M. M. (1872). *Treasury of Thought: Forming an Encyclopaedia of Quotations from Ancient and Modern Authors.* Boston: J. R. Osgood and Co., p. 433.

INDEX